the verbal math lesson 3

LEVEL 3
FOR CHILDREN AGES 8 TO 11

MICHAEL LEVIN, M.D.
CHARAN LANGTON, M.S.

Copyright 2008 Mountcastle Company
First Print Edition February 2008
First Electronic Edition June 2011
Second Edition April 2014

Edited by Kelsey Negherbon, Ashley Kuhre and Julie Lundy
Design by Tijana Mihajlović

Manufactured in the United States of America

ISBN 978-0-913063-29-3

www.mathlesson.com

INTRODUCTION

In math – more than in any other subject – a student progresses to the next level of math only after mastering all the previous levels. This step-by-step mastery is achieved through developing automaticity, which is the ability to solve problems with minimal effort. Automaticity is achieved through repeated exposure.

Typical elementary school curriculum uses written math as a path toward automaticity. While easy for some, the additional task of looking at the problem, copying it, and then writing down the solution hinders the acquisition of new skills and development of intuition. Excessive writing makes it cumbersome, eventually leading to an aversion towards math. Worksheets-type problems isolate math from real life. Many children have problems with word problems later in school, because they have not connected math to everyday life. Word problems in this program help to develop this connection.

The Verbal Math Lesson series recognizes and eliminates these barriers that block math proficiency. The program allows the child to direct her focus strictly on math and its algorithms. Many basic math facts need to be simply memorized and this program reinforces them in various ways.

Verbal Math is a useful skill for everyday life. It can be used in daily transactions, from figuring out the amount of change from $20 for lunch to estimating the number of miles a car can go on 5 gallons of fuel. We rely on our mental skills, not on the availability of paper, pencils, or a calculator.

The third book of the Verbal Math series follows the same step-by-step progression proved so successful in earlier volumes. You may consider revisiting previous books in the series if the child has difficulties with the problems here.

In the third book of the Verbal Math series, we reinforce addition and subtraction of double and triple-digit numbers, single and double-digit multiplication and division. We will present rates and ratios, teach factors, and introduce fractions. As always, we expect all new learning to be natural and fun.

Verbal Math is not a collection of simple problems designed to give the student practice in arithmetic, nor is it an Exercise book to accompany a written math course. Instead, verbal math has a separate set of rules that lead to mastery in a very essential skill - mental computations. Doing problems in your head leads to developing shortcuts and an intuitive understanding which is often overlooked in written math. Additionally, Verbal Math is an indispensable skill when the child moves to fractions, percentages, and, eventually, to algebra.

These lessons are built on learned and reinforced concepts. Each lesson follows a familiar plan: the explanation of a new concept, Exercise sections, and word problems. In the beginning of Word Problems sections we offer solutions with explanations of the procedures and results. If the child answers these problems correctly the first time, there is no need to go over the hints or solutions. In the Word Problems sections, ask the student to give the answer with the relevant units, not just the numbers.

Some word problems might be confusing. Before starting the calculations, the child needs to "blueprint" the problem in her mind to decide what we are measuring, what type of calculation applies, and what units will be used in the answer.

Chapter 10 (measuring and calculating time) might be a challenge for some children. The child must be able to tell time and understand the concept of 24 hours in a day, 60 minutes in an hour, and 60 seconds in a minute. If you encounter difficulties, skip this chapter for now and move on. This is a stand-alone lesson and there will be no interruption in continuity.

As always, please send us your comments and observations.
Thanks you, Best of luck,
Michael Levin and Charan Langton

www.mathlesson.com

BOOK THREE

PROPERTIES OF NUMBERS

Numbers you multiply together are called *factors* of the *product*. For example, in 3 × 2 = 6, number 3 and 2 are factors of 6. Also, in 1 × 6 = 6, we call 1 and 6 factors of 6.

We can say 1, 2, 3, 6 are all possible factors of number 6.

- Number 8 has factors 1, 2, 4, and 8 because both 1 × 8 and also 2 × 4 make 8.
- Number 9 has factors 1, 3, and 9.
- Number 10 has 1, 2, 5, and 10
- Number 21 has 1, 3, 7, and 21.

Factors are also numbers that divide the number without a remainder.

When a number has factors other than 1 and itself, we call it a *composite number*. If a number can be divided only by 1 and itself, we call this number *prime*.

- 1 is a prime number, divided by 1 and itself.
- 2 is a prime number, divided only by 1 and itself.
- 3 is a prime number, divided only by 1 and itself.

- 4 is a composite number, because in addition to 1 and itself, it can be divided by 2.
- 5 is prime number (factors 1 and 5)
- 6 is a composite number with factors 2 and 3, in addition to 1 and 6.
- 7 is a prime number (factors 1 and 7)
- 8 is a composite number (factors 1, 2, 4, and 8)
- 9 is a composite number (factors 1, 3, and 9)
- 10 is a composite number (factors 1, 2, 5, and 10)
- 11 is a prime number
- 12 is a composite number (factors 1, 2, 3, 4, 6, and 12)

EXERCISE I

1. Is 14 prime or composite?
   Ans: composite (factors 1, 2, 7, and 14)

2. Is 17 prime or composite? *Ans:* Prime

3. Is 19 prime or composite? *Ans:* Prime

4. Is 21 prime or composite?
   Ans: Composite (factors (1, 3, 7, and 21)

5. Is 23 prime or composite? *Ans:* Prime

6. Is 24 prime or composite?
   Ans: Composite (factors 1, 2, 3, 4, 6, 8, 12, and 24)

7. Is 25 prime or composite? *Ans:* Composite (factors 1, 5, and 25)

EXERCISE II

1. What are the factors of number 12? *Ans:* 1, 2, 3, 4, 6, 12.
   One and 12 are kind of obvious so we can skip them.

2. What are factors of number 8? *Ans:* 2 and 4.

3. What are factors of number 15? *Ans:* 3 and 5.

4. What are factors of number 14? *Ans:* 2 and 7.

5. What are factors of number 18? *Ans:* 2, 3, 6, 9.

6. What are factors of number 21? *Ans:* 3, 7.

7. What are factors of number 25? *Ans:* Just 5.

8. What are factors of number 23? *Ans:* None other than 23.
    It is a prime number.

9. What are factors of number 17? *Ans:* None other than 17.
    It is a prime number.

10. What are factors of number 27? *Ans:* 3, 9.

11. What are factors of number 33? *Ans:* 3, 11.

12. What numbers multiplied together give 24? Give all
    combinations. *Ans:* 4 × 6, also, 2 × 2 × 6, also 2 × 2 × 2 × 3.

13. What numbers multiplied together give 36? Give all
    combinations. *Ans:* 2 × 18, also, 2 × 3 × 6, also 2 × 2 × 3 × 3.

14. What numbers multiplied together give 34? Give all
    combinations. *Ans:* 2 × 17, that's all.

15. What numbers multiplied together give 40? Give all
    combinations. *Ans:* 2 × 20, also, 2 × 2 × 10, also 2 × 2 × 2 × 5.

16. What numbers multiplied together give 50? Give all
    combinations. *Ans:* 2 × 25, also, 2 × 5 × 5, 2 × 10 × 5.

17. What numbers multiplied together give 60? Give all
    combinations. *Ans:* 2 × 30, also, 2 × 2 × 15, 2 × 2 × 5 × 3.

18. What numbers multiplied together give 75? Give all
    combinations. *Ans:* 5 × 15, also, 3 × 25, 3 × 5 × 5.

19. What numbers multiplied together give 90? Give all
    combinations. *Ans:* 3 × 30, also, 3 × 2 × 15, 3 × 2 × 5 × 3,
    2 × 45, 2 × 15 × 3, 10 × 9, 10 × 3 × 3.

20. What numbers multiplied together give 100? Give all
    combinations. *Ans:* 2 × 50, also, 2 × 25, also, 2 × 5 × 5, 2 × 10 ×
    5, 25 × 4, 5 × 5 × 2 × 2, etc.

WORD PROBLEMS

1. How can we divide six kittens among 1, 2, 3, and 6 friends?
   Ans: 6 kittens, 3 kittens, 2 kittens, or 1 kitten each.
   Solution: One friend can keep all six kittens, that is 6 ÷ 1 = 6
   (kittens).Two friends can divide kittens between themselves:
   6 ÷ 2 = 3 (kittens).
   Three friends cad divide them three ways: 6 ÷ 3 = 2 (kittens).
   Six friends can do it too: 6 ÷ 6 = 1 (kitten for each friend)

2. A customer paid $16 for 2 mangosteens. What's the price of
   one mangosteen? *Ans:* $8.
   Mangosteen is a tropical fruit that grows in Indonesia. A legend
   says that English Queen, Victoria offered 100 pounds sterling
   (a lot of money in those days) to anyone who brings her a fresh
   mangosteen fruit. Some people still call the mangosteen the
   "Queen of Fruit".

3. A customer paid $16 for 4 melons. What's the price of one
   melon? *Ans:* $4.

4. A customer paid $16 for 8 grapefruits. What's the price of one
   grapefruit? *Ans:* $2.

5. If you divide 42 chores evenly among 7 kids, how many chores
   will each kid get? *Ans:* 6 chores each.
   a) What if there were only 6 kids? *Ans:* 7 chores.
   b) Does each kid get more or less chores if there are 3 kids?
   Ans: More chores.

6. How many stops will a bicyclist make on a 54 mile trip, if he is
   stopping every 9 miles? *Ans:* 5 stops, not counting the last one.

7. Arthur worked on a school project for 42 days. How many
   weeks is that? *Ans:* 6 weeks.

8. If one stapler costs $3, how many staplers can you buy for $15?
   Ans: 5 staplers.

9. One stock share costs $7. How many stock shares can you buy
   for $42? *Ans:* 6 stock shares. You can solve the problem without
   even knowing what a stock share is. Hooray!

10. If each gizmo needs 5 doohickeys, how many gizmos need 35 doohickeys? *Ans:* 7 gizmos.

11. After helping Mr. Pitts to pick pears, 6 helpers received 42 pounds of pears to divide among themselves. How many pounds did each get? *Ans:* 7 pounds.

12. How many times 4 is contained in 24? In 32? In 36? *Ans:* 6, 8, and 9.

13. How many times 6 is contained in 24? In 42? In 48? *Ans:* 4, 7, and 8.

14. How many times 7 is contained in 28? In 42? In 56? *Ans:* 4, 6, and 8.

15. A ticket to Buffalo costs $9. How many tickets can you buy for $54? *Ans:* 6 tickets.

16. If one ticket is $9, how much do 7 tickets cost? *Ans:* $63.

17. If one turkey has four legs, how many legs do 9 turkeys have? *Ans:* Ha-ha, turkeys have only 2 legs, but if they had 4, then it would be 36.

18. If 6 pencil holders cost $42, how much does one cost? *Ans:* $7

19. If 7 pencil holders together hold 56 pencils, how many pencils are in each holder? *Ans:* 8 pencils.

20. If 7 firefighters ride on one fire truck, how many firefighters ride on 9 trucks? *Ans:* 63 firefighters.

21. It takes 9 logs to make a raft.
    a) How many logs does it take to make 4 rafts? *Ans:* 36 logs.
    b) How many logs does it take to make 6 rafts? *Ans:* 54 logs.
    c) How many logs does it take to make 8 rafts? *Ans:* 72 logs.
    d) How many logs does it take to make 9 rafts? *Ans:* 81 logs.

22. If it takes 5 minutes to inflate 35 balloons, how many balloons can you inflate in one minute? *Ans:* 7 balloons.

23. If one artist uses 8 brushes to paint a landscape, how many brushes would 6 artists use? *Ans:* 48 brushes.

24. A rabbit can jump 7 feet in one jump. How many feet can it jump in 5 jumps? *Ans:* 35 feet.
    How many feet can you jump in 5 jumps?

25. A gold digger divided 54 ounces of gold into 6 bags. How many ounces of gold were in each bag? *Ans:* 9 ounces.

26. If 45 acres of land are divided into 5 equal parcels, how big is each parcel? *Ans:* 9 acres.
    a) If 7 oak trees are planted on each parcel, how many trees were planted on 5? *Ans:* 35 trees.

27. If a teaspoon holds 5 grams of water. How many grams of water are in 6 spoons? *Ans:* 30 grams. A gram is a very small weight. There are almost 30 grams in each ounce of water and 3,785 grams in each gallon.

28. If a tablespoon holds 15 grams of water, how many teaspoons is that? *Ans:* 3 teaspoons, because each teaspoon is 5 grams.

29. If a spider catches 8 flies in one week, how many flies will it catch in 4 weeks? *Ans:* 32 flies.

30. If a fly can catch 4 spiders in one week, how many spiders will it catch in 7 weeks? *Ans:* 28 spiders, as silly as that might be.

31. It takes Benjie 8 minutes to take a bath. For Angie, it takes 5 times longer. How long does it take Angie? *Ans:* 40 minutes.

32. A 24 hour work day was divided into 3 shifts. How long is each shift? *Ans:* 8 hours.

33. A square has 4 angles. How many squares have 24 angles?
    Ans: 6 squares.

34. A pentagon has 5 angles. How many pentagons have 35 angles?
    Ans: 7 pentagons.

35. If a hexagon has 6 angles. How many hexagons have 54 angles?
    Ans: 9 hexagons.

36. An octagon has 8 angles. How many angles will 8 octagons have?
    Ans: 64 angles.
    Do you know the name of a figure with 10 angles? It's decagon. And polygon is a figure with many angles, because poly means 'many' in Greek.

37. A postman delivered 63 newspapers and 7 times fewer magazines. How many magazines did he deliver?
   Ans: 9 magazines.

38. One astronaut circled the Earth 6 times, another 8 times more. How many times did the second astronaut circle the Earth?
   Ans: 48 times.

39. There are 6 ounces of tomato juice in one can. How many ounces are in 7 cans? *Ans:* 42 ounces.

40. If it takes 7 tomatoes to make one can of juice, how many tomatoes does it take to make 7 cans? *Ans:* 49 tomatoes.

41. A cat crawled 6 feet in the grass and then ran 6 times as many feet to catch a squirrel. How many feet did it crawl and run?
   Ans: 42 feet (6 feet crawling and 36 feet running).

42. If a team paints 6 walls in one day, how many walls can they paint in one week? *Ans:* 42 walls.

43. One printer prints 3 pages per minute. Another prints twice as fast. How many pages do both print in 8 minutes?
   Ans: 72 pages. The first printer will print 24 pages, the second 48. Together they'll print 24 + 48 = 72 (pages).

44. Making a retaining wall, a mason laid 9 bricks in one row. He needs to make 7 rows. How many total number of brick will he need? *Ans:* 63 bricks.

45. There are 9 holes in one slice of Swiss cheese. How many holes are in 6 slices? *Ans:* 54 holes.

46. If there are 7 vertebrae bones in one neck, how many bones are in 8 necks? *Ans:* 56 bones.

47. If Scott makes 8 spelling errors on every page, how many errors will he make on 9 pages? *Ans:* 72 errors.

48. How many squares are on a chess board, if there are 8 rows with 8 squares in each row? *Ans:* 64 squares.

49. Each costume in a school play uses 9 safety pins. How many pins will they need for 9 costumes? *Ans:* 81 pins.

50. If each battery has 9 volts, how many volts are in 5 batteries? *Ans:* 45 volts.

51. If each door has 5 hinges, how many hinges do 8 doors have? *Ans:* 40 hinges.

52. If each secret potion has 7 parts, how many parts are in 5 potions? *Ans:* 35 parts.

MULTIPLICATION OF SINGLE DIGITS AND DIVISION WITH NUMBERS UP TO 90

EXERCISE I

Check multiplication facts. Ask child to give the answer quickly.

$2 \times 9 = ?$ *Ans:* 18 $4 \times 7 = ?$ *Ans:* 28 $6 \times 8 = ?$ *Ans:* 48

$4 \times 4 = ?$ *Ans:* 16 $7 \times 5 = ?$ *Ans:* 35 $5 \times 9 = ?$ *Ans:* 45

$5 \times 5 = ?$ *Ans:* 25 $4 \times 9 = ?$ *Ans:* 36 $6 \times 7 = ?$ *Ans:* 42

$8 \times 3 = ?$ *Ans:* 24 $7 \times 6 = ?$ *Ans:* 42 $6 \times 9 = ?$ *Ans:* 54

$4 \times 6 = ?$ *Ans:* 24 $8 \times 4 = ?$ *Ans:* 32 $8 \times 7 = ?$ *Ans:* 56

$7 \times 3 = ?$ *Ans:* 21 $5 \times 6 = ?$ *Ans:* 30 $5 \times 8 = ?$ *Ans:* 40

$4 \times 5 = ?$ *Ans:* 20 $6 \times 4 = ?$ *Ans:* 24 $8 \times 9 = ?$ *Ans:* 72

$3 \times 9 = ?$ *Ans:* 27 $7 \times 7 = ?$ *Ans:* 49 $9 \times 9 = ?$ *Ans:* 81

EXERCISE II

$12 \div 2 = ?$ *Ans:* 6 $16 \div 8 = ?$ *Ans:* 2 $36 \div 4 = ?$ *Ans:* 9

$12 \div 3 = ?$ *Ans:* 4 $32 \div 4 = ?$ *Ans:* 8 $25 \div 5 = ?$ *Ans:* 5

$12 \div 4 = ?$ *Ans:* 3 $24 \div 3 = ?$ *Ans:* 8 $18 \div 3 = ?$ *Ans:* 6

$12 \div 6 = ?$ *Ans:* 2 $32 \div 8 = ?$ *Ans:* 4 $36 \div 6 = ?$ *Ans:* 6

$16 \div 2 = ?$ *Ans:* 8 $27 \div 3 = ?$ *Ans:* 9 $42 \div 6 = ?$ *Ans:* 7

$16 \div 4 = ?$ *Ans:* 4 $28 \div 4 = ?$ *Ans:* 7 $45 \div 5 = ?$ *Ans:* 9

$49 \div 7 = ?$ *Ans:* 7 $56 \div 8 = ?$ *Ans:* 7 $73 \div 9 = ?$ *Ans:* 7

$36 \div 6 = ?$ *Ans:* 6 $63 \div 9 = ?$ *Ans:* 7 $81 \div 9 = ?$ *Ans:* 9

$54 \div 6 = ?$ *Ans:* 9 $45 \div 9 = ?$ *Ans:* 5 $42 \div 6 = ?$ *Ans:* 6

$42 \div 7 = ?$ *Ans:* 6 $64 \div 8 = ?$ *Ans:* 8 $57 \div 7 = ?$ *Ans:* 8

EXERCISE III

1. What number do you multiply by 2 to get 18? *Ans:* 9

2. What number do you multiply by 7 to get 21? *Ans:* 3

3. What number do you multiply by 4 to get 24? *Ans:* 6

4. What number do you multiply by 5 to get 30? *Ans:* 6

5. What number do you multiply by 6 to get 36? *Ans:* 6

6. What number do you multiply by 8 to get 40? *Ans:* 5

7. What number do you multiply by 4 to get 36? *Ans:* 9

8. What number do you multiply by 9 to get 54? *Ans:* 6

9. What number do you multiply by 7 to get 49? *Ans:* 9

10. What number do you multiply by 8 to get 72? *Ans:* 9

11. Name all single-digit numbers by which 12 can be divided?
    Ans: 1, 2, 3, 4, 6

12. Name all single-digit numbers by which 9 can be divided?
    Ans: 1, 3

13. Name all single-digit numbers by which 24 can be divided?
    Ans: 1, 2, 3, 4, 6, 8

14. Name all single-digit numbers by which 36 can be divided?
    Ans: 1, 2, 3, 4, 6, 9

15. Name all single-digit numbers by which 49 can be divided?
    Ans: 1, 7

EXERCISE IV

In this exercise, we will ask you to name all two *single-digit* pairs that multiplied together make the number, except for 1 and the number itself.

- Can you name all single-digit factors of 8 other than 1 and 8?
  Ans: Factors 2 and 4. When multiplied they make 8.
- Can you name all single-digit factors of 12?
  Ans: Number 12 has two pairs of single-digit factors:
  2 and 4 make a pair, and also 3 and 4.

1. Name all single-digit factor pairs to make 14? *Ans:* 2 and 7

2. Name all single-digit factor pairs to make 16? *Ans:* 2 and 8 and also 4 and 4

3. Name all single-digit factor pairs to make 18? *Ans:* 2 and 9 and also 3 and 6

4. Name all single-digit factor pairs to make 20? *Ans:* 4 and 5

5. Name all single-digit factor pairs to make 28? *Ans:* 4 and 7

6. Name all single-digit factor pairs to make 21? *Ans:* 3 and 7

7. Name all single-digit factor pairs to make 24? *Ans:* 4 and 6 and also 3 and 8

8. Name all single-digit factor pairs to make 27? *Ans:* 3 and 9

9. Name all single-digit factor pairs to make 42? *Ans:* 6 and 7

10. Name all single-digit factors pairs to make 49? *Ans:* 7 and 7

11. Name all single-digit factors pairs to make 36? *Ans:* 4 and 9 and also 6 and 6

12. Name all single-digit factors pairs to make 56? *Ans:* 8 and 7

13. Name all single-digit factors pairs to make 54? *Ans:* 6 and 9

3

COMMON DIVISOR AND GREATEST COMMON DIVISOR

Two or more numbers may have a *common divisor*, or the same number(s) by which we can divide all of them. For example, the numbers 4 and 8 have two common divisors: 2 and 4. Usually, we don't call "1" a common divisor.

The *greatest common divisor* is the largest number by which you can divide two or more numbers. For example, numbers 12 and 18 have several common divisors: 2, 3, and 6. But only 6 is the greatest common divisor for these numbers.

- The numbers 10 and 15 have 5 as their common divisor.
- The numbers 14 and 21 have 7 as a common divisor.
- The numbers 18 and 27 have 9 as common divisor and also 3.
- The numbers 30 and 50 have 5 and 10 as common divisors and also 2.
- The numbers 24 and 32 have 2, 3, 4, and 8 as common divisors.

1. What are common divisors for 8 and 10? ***Ans:*** 2
2. What are common divisors for 6 and 9? ***Ans:*** 3

3. What are common divisors for 12 and 15? *Ans:* 3

4. What are common divisors for 20 and 25? *Ans:* 5

5. What are common divisors for 15 and 18? *Ans:* 3

6. What are common divisors for 15 and 16? *Ans:* None

7. What are common divisors for 7 and 14? *Ans:* 7

8. What are common divisors for 21 and 27? *Ans:* 3

9. What are common divisors for 21 and 25? *Ans:* None

10. What are common divisors for 10, 20 and 30?
    Ans: 10 (and also 2)

11. What are common divisors for 6, 18, 24? *Ans:* 2, 3, and 6

WORD PROBLEMS

1. There were 16 parrots in 3 cages. The zookeeper added 2 parrots to each cage. How many parrots are in all cages now? *Ans:* 22 parrots.
   Solution: He added 2 (parrots) × 3 (cages) = 6 (parrots). Now, there are 16 + 6 = 22 (parrots). Don't get tricked, the problem didn't ask how the parrots were divided among three cages.

2. Kumar received 5 cards for his high school graduation and 4 times as many for his birthday. How many cards did he receive? *Ans:* 25 cards (5 × 4 = 20; 5 + 20 = 25)

3. There were 4 boxes with 6 books in each box. Then, the librarian brought 2 more boxes. How many books are there now? *Ans:* 36 books.
   Solutions: There are two ways to solve the problem.
   a) There were 6 (books) × 4 (boxes) = 24 (books) at first. They added 6 (books) × 2 (boxes) = 12 (books). Now there are 24 + 12 = 36 (books).
   b) Better way. There were 4 boxes, now there are 4 + 2 = 6 (boxes). If there are 6 books in each box, then 6 (books) × 6 (boxes) = 36 (books).

4. What are two *single-digit* factors of 42? *Ans:* 6 and 7.

5. There were 5 carrot patches with 8 carrots in each patch. Mrs. Bunny pulled out 17 carrots. How many carrots were left? **Ans:** 23 carrots (8 × 5 = 40, then 40 - 17 = 23).

6. I saw 7 rows of vehicles in a parking lot, with 7 in each row. There were 18 trucks, the rest were cars. How many cars were on the lot? **Ans:** 31 cars (7 × 7 = 49; then, 49 - 18 = 31).

7. There are 8 school busses in the parking lot. Each bus has 6 wheels. How many wheels do all the busses have? **Ans:** 48 wheels.

8. If a snail moves 3 inches in one minute, how far can it move in 9 minutes? **Ans:** 27 inches.
    To help it to move, the snail produces slime which protects the snail and helps it to glide over sharp objects without being injured. If you are a snail, you love your slime.

9. Ginger counted 6 helicopters and 4 times as many airplanes in an airfield. How many helicopters and airplanes together did she count? **Ans:** 30 aircraft.

10. Which two numbers when multiplied make 28? **Ans:** 4 and 7, and also 2 and 14. You are also right if you said 1 and 28, but that would be too easy.

11. There are 5 red flowers in the vase and 4 times as many white flowers. How many flowers are in the vase? **Ans:** 25 flowers.

12. In the showroom, there are 3 sofas with 5 pillows on each and 6 sofas with 4 pillows on each. How many pillows are there? **Ans:** 39 pillows.
    Solution: 5 pillows × 3 (sofas) = 15 pillows; 4 pillows × 6 (sofas) = 24 pillows. Then, 15 + 24 = 39 (pillows).

13. After putting 6 pictures in each of the 8 envelopes, Omar found 17 more pictures on the desk. How many pictures were there altogether? **Ans:** 65 pictures (6 × 8 = 48, 48 + 17 = 65).

14. Latisha lost 4 chess games but won 9 times as many. How many games did she play if there were no draws? **Ans:** 40 games.

15. Felipe drew 6 pictures a day for 7 days. Gino drew 11 pictures less over the same amount of time. How many pictures did both draw together?
    Ans: 73 pictures (6 × 7 = 42, 42 - 11 = 31; 42 + 31 = 73).

16. There were 36 berries on a plate. Jodie took 9 and divided the rest among her 3 cousins. Did she take more or less berries than she gave to each cousin?
    Ans: the same, she took 9 and gave away 9 berries to each cousin (36 - 9 = 27; 27 ÷ 3 = 9).

17. Mr. Seed picked 7 apples from each of 5 apple trees and 19 more apples on the ground. How many apples did he pick?
    Ans: 54 apples.

18. What number do you multiply by 3 to get 27? *Ans:* 9

19. In the park a forest ranger marked 6 birches and 7 times as many oaks. How many trees did she mark? *Ans:* 48 trees.

20. In a stable there are 9 race horses and 8 jockeys (people who ride the horses). How many legs do the horses and the jockeys have altogether? *Ans:* 52 legs.
    Solution: The horses have 4 legs × 9 (horses) = 36 legs; the jockeys (people) have 2 legs × 8 (jockeys) = 16 legs. Altogether they have 36 + 16 = 52 (legs), even though it's silly to count horses' and people's legs together.

21. Little Lynn is 5 years old. Her sister is 3 times older, and her father is 3 times older than her sister. How old is Lynn's father?
    Ans: 45 years old.
    Solution: If Lynn's sister is 3 times older, then she is 5 × 3 = 15 (years old). If Lynn's father is 3 times older than her sister, then he is 15 × 3 = 45 (years). You'll learn to multiply two-digit numbers later on; meanwhile, add 15 three times: 15 + 15 + 15 = 45.

22. There are 8 bicycles and 9 tricycles in a store. How many wheels are there in all?
    Ans: 34 wheels (2 × 8 = 16; 3 × 9 = 18; 16 + 18 = 34).

23. What's the greatest *common divisor* for 24 and 42?
    Ans: 6, it's also the only one.

24. There are 4 spiders and 7 flies on the old window. How many legs do these critters have together?
    Ans: 74 legs (8 × 4 = 32, 6 × 7 = 42; 32 + 42 = 74). Imagine spiders and flies sitting together and counting their legs.

25. Rita picked up 7 shells and 7 times as many pebbles on the beach. How many of both did she pick up? **Ans:** 56 shells and pebbles.

26. There were 43 books on 4 shelves. The kids took 3 books from each shelf. How many books are on all shelves now?
    Ans: 31 books
    Solution: They took 3 (books) × 4 (shelves) = 12 (books removed); then 43 - 12 = 31 (books left on the shelves).

27. What's the greatest common divisor for 17 and 56?
    Ans: There are none.

28. Give examples of 2 numbers with one divided by another making 6? **Ans:** 12 and 2; 18 and 3; 24 and 4; 30 and 5; etc.

29. A bakery made 53 pastries. They sold 11 pastries to a restaurant and the rest divided equally among 7 customers. How many did each customer get? **Ans:** 6 pastries (53 - 11 = 42; 42 ÷ 7 = 6).

30. There were 70 gallons of water in the barrel. Morris spilled 25 gallons and divided the rest equally into 9 cans. How many gallons did he pour in each can? **Ans:** 5 gallons (70 - 25 = 45; 45 ÷ 9 = 5).

31. Tyron spent 5 vacation weeks with his grandparents in the city and 3 weeks with his parents at the beach. How long in days was Tyron's vacation? **Ans:** 56 days.
    Solution: Tyron's vacation lasted 5 + 3 = 8 (weeks); 7 days (each week) × 8 (weeks) = 56 days.

32. There are 9 books with hard cover on the bookcase and 9 times as many paperbacks. How many books are on the shelf?
    Ans: 90 books

33. One shop fixed 8 cars a day for 5 days. Another shop fixed 13 more cars than the first. How many cars did both fix?
    Ans: 93 cars (8 × 5 = 40; 40 + 13 = 53; both: 40 + 53 = 93).

34. School cafeteria bought 7 knives, twice as many spoons, and 4 times as many forks. How many utensils did they buy? ***Ans:*** 49 utensils (7 + 14 + 28 = 49).

35. There were 8 jars with jam on the table. Oscar took 3 spoonfuls of jam from each jar. How many spoonfuls of jam did he take? ***Ans:*** 24 spoons.

36. A restaurant set 4 tables with 4 plates on each table, 5 tables with 5 plates on each, and 6 tables with 6 plates on each.
a) How many plates were there? ***Ans:*** 77 plates (16 + 25 + 36 = 77)
b) Only 24 plates were used. How many weren't? ***Ans:*** 53 plates.

37. The first book has 8 pictures, the other 3 times as many, and the third book has 20 pictures. How many pictures are in all three books? ***Ans:*** 52 pictures.

38. If a tall house has 8 windows and a short house has only 4, how many windows are in 5 tall and 5 short houses together? ***Ans:*** 60 windows.

39. Which two numbers when multiplied make 32? ***Ans:*** 4 and 8, if you answered 2 and 16 it's also right; 1 and 32 is also correct.

40. A building has nine 4-bedroom apartments and seven 3-bedrooms. How many bedrooms are in the building? ***Ans:*** 57 bedrooms (36 + 21 = 57).

41. A chessboard has 8 rows of squares. Bobby Fischer put 4 chess pieces in each row. How many pieces did he put out? ***Ans:*** 32 pieces. Did you know that the game of chess came from India and is one of the oldest games?

42. Mark took some olives from a jar and put them on 7 plates, with 5 olives on each plate. If there are 15 olives left in the jar, how many were there at first? ***Ans:*** 50 olives.

43. There were 9 bad days during the trip and 8 times as many good days. How many days did the trip last? ***Ans:*** 81 days.

44. After 95 new recruits came to the regiment, 38 were sent away to another unit, 1 went home, and the rest were divided into 7 equal groups. How many recruits were in each group? ***Ans:*** 8 recruits (95 - 38 = 57, 57 - 1 = 56; 56 ÷ 7 = 8), the one who left probably missed his mommy.

45. A dry cleaner received 5 jackets and 10 times as many pants. How many items did they receive? *Ans:* 55 items.

46. Lola built 2 shelves and put 3 vases on each shelf. Then she built another shelf and divided her vases evenly among them. How many vases are on each shelf now? *Ans:* 2 vases.
Solution: Lola had 3 vases × 2 shelves, that's 6 (vases). She made another shelf and divided 6 vases among 3 shelves, that's 6 ÷ 3 = 2 (vases on each shelf).

47. Diana had 5 fish tanks with 4 fish in each tank. While cleaning she broke one tank and had to divide fish evenly among the other tanks. How many fish are in each tank now? *Ans:* 5 fish (5 × 4 = 20; 20 ÷ 4 = 5).

48. Monica kept 9 stamps in each of her 6 albums. She bought two new stamps and one new album and decided to divide all her stamps evenly again. How many stumps are in each album now? *Ans:* 8 stamps.
Solution: Monica had altogether 9 (stamps) × 6 (albums) = 54 stamps. She bought 2 new stumps, 54 + 2 = 56. She bought one more album decided to divide all her stamps into 7 albums; that's 56 (stamps) ÷ 7 = 8 stamps in each album.

49. I am thinking of a number. I can divide my number by 2, 3, 4, or 6.
a) What's my number? *Ans:* 12
b) Can you think of other numbers that can do that?
Ans: 24, 48, and others.

50. I have a number of bricks that is less than 30. I can divide all my bricks into either 3 equal stacks or 9 equal stacks.
How many bricks do I have?
Ans: 27 bricks. You can't say 9 bricks because one brick doesn't make a stack.

51. They brought 8 benches to seat 6 kids on each bench. Now, two benches are empty and there are equal number of kids on the rest of the benches. How many children sat on each bench?
Ans: 8 children.
Solution: They expected 6 kids on each bench, that's 6 (kids) × 8 (benches) = 48 children. The children are sitting now on 6 benches. Then, 48 ÷ 6 = 8 (children on each bench).

ADDING NUMBERS THAT END IN 0, WITH SUMS OVER 100

You already know how to add double digit numbers less than 100. Adding double digit numbers with sum over 100 is not hard either.

Problem: 90 + 20 = ?
Solution: 20 is made if two 10's, 10 + 10 = 20.
Then, 90 + 10 = 100 and 100 + 10 = 110.
Hence, 90 + 20 = 110

Problem: 70 + 50 = ?
Solution: It takes 30 to add to 70 to make 100, We know that 50 splits into 30 + 20.
Then, 70 + 30 = 100 and 100 + 20 = 120.

Now add these numbers quickly. Look at the left most number first.

70 + 50 = ? *Ans:* 120	90 + 30 = ? *Ans:* 120	80 + 80 = ? *Ans:* 160
90 + 10 = ? *Ans:* 100	60 + 60 = ? *Ans:* 120	90 + 60 = ? *Ans:* 150
90 + 20 = ? *Ans:* 110	40 + 90 = ? *Ans:* 130	90 + 70 = ? *Ans:* 160
30 + 70 = ? *Ans:* 100	70 + 70 = ? *Ans:* 140	70 + 80 = ? *Ans:* 150
30 + 80 = ? *Ans:* 110	50 + 70 = ? *Ans:* 120	90 + 90 = ? *Ans:* 180
50 + 60 = ? *Ans:* 110	90 + 50 = ? *Ans:* 140	90 + 20 = ? *Ans:* 110

Adding triple-digit numbers ending with 0. First, we add numbers in hundreds place. Next, we add numbers in tens place and then, put together two sums.

Problem: 120 + 110 = ?
Solution: 120 + 100 = 220, then add 10 more, and we get 230.

Problem: 170 + 140 = ?
Solution: 170 + 100 = 270, then add 40 more and we get 310.

EXERCISE I

50 + 60 = ? *Ans:* 110	90 + 90 = ? *Ans:* 180	190 + 30 = ? *Ans:* 220
90 + 40 = ? *Ans:* 130	100 + 50 = ? *Ans:* 150	200 + 50 = ? *Ans:* 250
70 + 50 = ? *Ans:* 120	100 + 90 = ? *Ans:* 190	180 + 60 = ? *Ans:* 240
80 + 60 = ? *Ans:* 140	110 + 30 = ? *Ans:* 140	170 + 80 = ? *Ans:* 250
70 + 70 = ? *Ans:* 140	110 + 70 = ? *Ans:* 180	240 + 50 = ? *Ans:* 290
80 + 70 = ? *Ans:* 150	120 + 80 = ? *Ans:* 200	260 + 40 = ? *Ans:* 300
80 + 80 = ? *Ans:* 160	150 + 50 = ? *Ans:* 200	270 +40 = ? *Ans:* 310
90 + 50 = ? *Ans:* 140	140 + 60 = ? *Ans:* 200	250 + 150 = ? *Ans:* 400
30 + 80 = ? *Ans:* 110	190 + 10 = ? *Ans:* 200	190 + 20 = ? *Ans:* 210
70 + 90 = ? *Ans:* 160	190 + 20 = ? *Ans:* 210	190 + 40 = ? *Ans:* 230

EXERCISE II

90 + 100 = ? *Ans:* 290	180 + 150 = ? *Ans:* 330
160 + 10 = ? *Ans:* 170	190 + 150 = ? *Ans:* 340
160 + 110 = ? *Ans:* 270	230 + 60 = ? *Ans:* 290
120 + 120 = ? *Ans:* 240	110 + 240 = ? *Ans:* 350
130 + 120 = ? *Ans:* 250	320 + 160 = ? *Ans:* 480
250 + 50 = ? *Ans:* 300	250 + 220 = ? *Ans:* 470
130 + 140 = ? *Ans:* 270	240 + 250 = ? *Ans:* 490
160 + 130 = ? *Ans:* 290	140 + 90 = ? *Ans:* 230
150 + 150 = ? *Ans:* 300	240 + 90 = ? *Ans:* 330
170 + 130 = ? *Ans:* 300	220 + 190 = ? *Ans:* 410
190 + 110 = ? *Ans:* 300	120 + 320 = ? *Ans:* 440
190 + 120 = ? *Ans:* 310	160 + 230 = ? *Ans:* 390
170 + 150 = ? *Ans:* 320	330 + 150 = ? *Ans:* 480
160 + 140 = ? *Ans:* 300	360 + 240 = ? *Ans:* 600

370 + 280 = ? *Ans:* 650

290 + 210 = ? *Ans:* 500

170 + 80 = ? *Ans:* 250

130 + 180 = ? *Ans:* 310

220 + 190 = ? *Ans:* 410

WORD PROBLEMS

1. One movie theater has 60 seats, another has 80. How many seats are in both? *Ans:* 140 seats.

2. A ribbon was cut in the middle, and each piece measured 70 inches. How long was the ribbon before it was cut? *Ans:* 140 inches.

3. After 50 cars left the parking lot, there were 80 cars still there. How many cars were on the lot at first? *Ans:* 130 cars.

4. There are 60 one-story and 60 two-story houses in our town. How many houses are there altogether? *Ans:* 120 houses.

5. The home team scored 80 points and the guest team scored 70. How many points did both teams score? *Ans:* 150 points, but our team won!

6. One boxer punched the other 50 times during the boxing match. The other boxer punched the first 60 times. How many punches did were exchanged? *Ans:* 110 punches.

7. A kitchen drawer holds 50 forks, 50 spoons, and 50 knives. How many utensils are in the drawer? *Ans:* 150 utensils.

8. Mr. Dapper trimmed 60 hairs on his right mustache and 60 on his left. How many hairs did he trim? *Ans:* 120 hairs.

9. One monkey ate 60 bananas, the other ate 30 more than the first. How many bananas did they both eat? *Ans:* 150 bananas. (60 + 30 = 90, then 60 + 90 = 150)

10. There are 80 miles between point A and point B; and 90 miles between point B and point C. How many miles are between point A and point C? *Ans:* 170 miles.

11. Tuesday, a museum had 90 visitors. Wednesday, there were 110. How many visitors came on both days? *Ans:* 200 visitors.

12. One peacock has 130 "eyed" tail feathers, another has 90. How

many "eyed" tail feathers do both have? *Ans:* 210 feathers. Did you know that only male peacocks have colorful feather tails? These feathers grow to be several feet long and are shed each year. The feathers have a design near its tip which resembles an eye.

13. A bakery baked 120 poppy seed bagels and 110 sesame seeds. How many bagels did they bake? *Ans:* 230 bagels.

14. To impress his friends, Sam walked 70 feet like a duck, then hopped 80 feet like a rabbit, and then crawled another 40 feet like a snake. How many feet did he cover? *Ans:* 190 feet. Do you think he impressed his friends?

15. I put 140 sheets of paper into a printer and then added 130 more. How many sheets did I load? *Ans:* 270 sheets.

16. There are 150 seats on each side of the isle. How many seats are on both sides? *Ans:* 300 seats.

17. An airline pilot weighs 170 pounds, her co-pilot weighs 210 pounds. How much do both weigh? *Ans:* 380 pounds.

18. One number is 160 the other number 150. What's the sum? *Ans:* 310.

19. There were 120 passengers on the first train and 90 on the other. How many passengers were on both trains? *Ans:* 210 passengers.

20. There were 210 suitcases on the first train and 180 on the other. How many suitcases were on both trains? *Ans:* 390 suitcases.

21. There are 130 books in the first bin and 180 on the other. How many books are in both bins? *Ans:* 310 books.

22. One truck driver paid $130 to fill the tank with gas, the other paid $120, and the third paid $90. How much money did all three pay? *Ans:* $340.

23. Before noon, a food bank received 140 cans of food; after lunch they received 180 cans. How many cans did the bank get? *Ans:* 320 cans.

24. Paige collects picture postcards. She has 170 cards with flowers

and 250 cards with animals on them. How many cards of both kinds does she have? *Ans:* 420 cards.

25. If the first number is 100 and the second number is bigger by 100, and the third number is bigger than the second by 100, what is the sum of all three? *Ans:* 600
    Solution: The first number is 100; the second number is 100 + 100 = 200; the third number is 200 + 100 = 300. The sum is 100 + 200 + 300 = 600.

26. This year there were 270 sunny days and 40 rainy days. How many sunny and rainy days were in the year? *Ans:* 310 days. The rest of the days must have been overcast or snow.

27. From a long piece of fabric a tailor made 120 panama hats and 120 baseball hats. How many hats did she make? *Ans:* 240 hats.

28. Add 170 to 180. *Ans:* 350

29. The first day a county fair had 180 visitors, the second day 150 people. How many people visited? *Ans:* 330 people.

30. One day a truck drove 140 miles, the next day it drove 30 miles more than the first day. How many miles did the truck drive in two days? *Ans:* 310 miles (140 + 30 = 170, then 140 + 170 = 310).

31. Sleeping Beauty slept 220 days, then she stretched and slept 230 days more. How many days did she sleep? *Ans:* 450 days. She'll have lots of makeup homework to do when she wakes up.

32. First day on a new job, Kip delivered 140 flowers to the right addresses and 80 flowers to the wrong addresses. How many flowers did he try to deliver? *Ans:* 220 flowers.

33. A secret society kept 350 secrets and recently added 90 more secrets. How many secrets do they have now? *Ans:* 440 secrets.

34. If there are 320 miles from Los Angeles to San Francisco, how many miles are there round trip? *Ans:* 640 miles.

35. If a power boat has 230 horsepower in each of its two engines,

how many horsepower are in both? **Ans:** 460 horsepower, don't look for the horses inside, there are none.

36. A farmer picked 190 strawberries from one field and 380 from the other. How many strawberries did he pick? **Ans:** 570 berries.

37. One work team paved 580 square feet of the road, the other paved only 320. How much road did both teams pave? **Ans:** 900 square feet.

38. There are three notebooks, 120 pages in each book. How many pages are in all three? **Ans:** 360 pages.

39. There are three cans, 130 peas in each can. How many peas are in all three? **Ans:** 390 peas.

40. There are three piggy banks with 150 coins in each bank. How many coins are in all three? **Ans:** 450 coins.

41. There are three rooms in a local museum with 160 items in each room. How many items are in the museum? **Ans:** 480 items.

42. There are three witches; each witch knows 170 magic recipes. How many recipes do all three know? **Ans:** If you said 510 you counted well. Sometimes the witches share the recipes. Everyone has to share.

43. There are three apartment buildings with 180 apartments in each building. How many apartments are in all three? **Ans:** 540 apartments.

44. There are three hills, each 190 feet tall. How tall would a hill made of all three hills, one on top of the other? **Ans:** 570 feet.

SUBTRACTING FROM NUMBERS ENDING IN 0 OVER 100

There are several ways of doing verbal subtraction. We may need to split one or both numbers (the number we subtract from and the number we are subtracting) to make our calculations easier.

Problem: 160 - 40 = ?

Solution: 160 is made of 100 and 60. Next, we take 40 from 60, that's 20.

Then, we put 100 and 20 back together.

160 - 40 = 120

Problem: 120 - 50 = ?

Solution: 50 is equal to 20 + 30. First we will subtract 20 from 120, 120 - 20 = 100. Then, from 100 we subtract 30, 100 - 30 = 70.

120 - 50 = 70

Problem: 350 - 110 = ?

Solution: 110 is equal to 100 + 10. Next, 350 - 100 = 250, then 250 - 10 = 240.

350 - 110 = 240

Problem: 300 - 170 = ?

Solution: 170 = 100 + 70. Next, 300 - 100 = 200.
Then, 200 - 70 = 130.

300 - 170 = 130.

EXERCISE I

120 + 40 = 160	140 + 170 = 310	280 + 70 = 350
50 + 250 = 300	150 + 170 = 320	150 + 80 = 230
160 + 60 = 220	60 + 190 = 250	480 + 60 = 540
70 + 170 = 240	90 + 150 = 240	180 + 50 = 230
190 + 190 = 380	410 + 90 = 500	270 + 70 = 340
80 +1 20 = 200	150 + 50 = 200	530 + 90 = 620
80 + 130 = 210	160 + 50 = 210	350 + 90 = 440
180 + 250 = 430	160 + 70 = 230	770 + 70 = 840

EXERCISE II

120 + 40 = 160	160 + 50 = 210	600 - 20 = 580
50 + 250 = 300	120 - 20 = 100	700 - 80 = 620
160 + 60 = 220	130 - 20 = 110	500 - 70 = 430
70 + 170 = 240	150 - 40 = 110	160 + 70 = 230
190 + 190 = 380	200 - 30 = 170	280 + 70 = 350
80 +1 20 = 200	180 - 50 = 130	150 + 80 = 230
80 + 130 = 210	200 - 90 = 110	480 + 60 = 540
180 + 250 = 430	400 - 80 = 320	180 + 50 = 230
140 + 170 = 310	290 - 80 = 210	270 + 70 = 340
150 + 170 = 320	280 - 50 = 230	530 + 90 = 620
60 + 190 = 250	300 - 60 = 240	350 + 90 = 440
90 + 150 = 240	270 - 40 = 230	770 + 70 = 840
410 + 90 = 500	500 - 90 = 410	
150 + 50 = 200	200 - 70 = 130	

EXERCISE III

120 - 30 = 90	200 - 70 = 130	200 - 150 = 50
180 - 90 = 90	210 - 90 = 120	120 - 30 = 90
170 - 80 = 90	200 - 50 = 150	170 - 90 = 80

300 - 150 = 150	240 - 70 = 170	500 - 320 = 180
300 - 200 = 100	260 - 150 = 110	120 - 90 = 30
220 - 110 = 110	330 - 110 = 220	410 - 250 = 160
220 - 80 = 140	390 - 220 = 170	340 - 190 = 150

WORD PROBLEMS

1. One camp put 120 tents, and the other 30 tents less. How many tents are in the second camp? *Ans:* 90 tents.
    a) How many tents are in both camps? *Ans:* 210 tents

2. There should be 150 silver foxes in the park. The rangers counted 90. How many foxes are hiding? *Ans:* 60 foxes. Silver foxes are close cousins of common red foxes. Their fur is not silver but grey. People learned to tame and keep them as pets.

3. There were 400 trees in the forest and a tornado broke 70 trees. How many trees are still standing? *Ans:* 330 trees.

4. There were 130 airplanes on the ground and now there are only 50. How many airplanes took off? *Ans:* 80 airplanes.

5. There were 200 sheets of paper in the stack before Farisa took out 70. How many sheets are there now? *Ans:* 130 sheets.

6. There were 300 tin soldiers standing on the floor; a kitten ran and knocked down 90. How many are still standing? *Ans:* 210 tin soldiers.

7. A box had 200 pieces of tissue paper. Lora used 110 for her big sneeze. How many are left in the box? *Ans:* 90 pieces.

8. There are 200 trees in the grove, but 80 less in the square. How many trees are in the square? *Ans:* 120 trees.

9. Out of 240 trees inspected, the rangers marked 120 as healthy. How many trees needed treatment? *Ans:* 120 trees.
    Like people and animals, trees get sick. One of the worst trees' sicknesses is Dutch elm disease that destroyed half of the elm trees in the United States in the last 80 years.

10. There were 300 cans in the store and 160 were sold. How many cans are there now? *Ans:* 140 cans.

11. After bringing in 170 new rabbits, there are 380 endangered rabbits in the sanctuary now. How many rabbits were there before the addition? ***Ans:*** 210 rabbits.
    Solution: There are 380 rabbits now. Before there were 380 - 170 (new rabbits) = 210 rabbits (before the addition).
    Have you ever heard of an endangered rabbit? Volcano rabbits live in Mexico. They are tiny, have small rounded ears and short, thick fur. Unlike other rabbits, the volcano rabbit makes very high-pitched sounds instead of thumping its feet on the ground to warn other rabbits of danger.

12. Colin went up 110 steps. How many more does he have to climb, if there are 260 steps to the bottom to the top? ***Ans:*** 150 steps.

13. There were 300 plums on a tree and the squirrels ate 180 of them. How many plums were left for humans and other critters? ***Ans:*** 120 plums.

14. The laundromat received 240 shirts and washed 150 the first day. How many shirts are left to wash? ***Ans:*** 90 shirts.

15. The same laundromat received 350 bed sheets and washed 160 the first day. How many sheets are left to wash? ***Ans:*** 190 sheets.

16. After an argument, 180 bees moved out of the beehive. How many bees stayed if there were 350 bees before the quarrel? ***Ans:*** 170 bees. The quarrel created a buzz.

17. Alfred cut 190 feet from a 260 foot wire. How much wire was left after the cut? ***Ans:*** 70 feet.

18. An earthworm crawled 250 inches in two days. How many inches did the worm crawl the second day if the first day it crawled 120 inches? ***Ans:*** 130 inches.

19. A museum moved 180 paintings to the new wing. How many paintings were left in the old building if the museum has 400 paintings total? ***Ans:*** 220 paintings.

20. To make two bracelets, a jeweler took 230 grams of silver and used 170 grams on one. How much silver was left for the second bracelet? ***Ans:*** 60 grams.

21. The children took out 180 jelly beans from a bag of 400. How many jelly beans did they leave in the bag? *Ans:* 220 beans. Why leave any jelly beans in the bag?

22. A zookeeper divided 300 fish between the penguins and dolphins. How many fish did the dolphins get if the penguins got 130? *Ans:* 170 fish.

23. Lara used 430 beads for two necklaces. If the first necklace was made of 190 beads, how many beads did she use for the other? *Ans:* 240 beads.

24. When the first necklace broke, 110 out of 190 beads fell on the floor. How many beads stayed on the string? *Ans:* 80 beads.

25. If 510 passengers took two trains and the first train had 280 passengers, how many were on the second train? *Ans:* 230 passengers.

26. After 350 newspapers were divided into two piles, how many newspapers were in the first pile, if the second had 180? *Ans:* 170 newspapers.

27. Paolo counted 230 calves on the field. How many bull calves did he count, if there were 110 cow calves? *Ans:* 120 bull calves.

28. Saturday, 390 visitors came to the exhibition. Sunday, there were 160 less visitors. How many people came Sunday? *Ans:* 230 people.

29. There are 440 miles between Boston and Washington, DC. If Dad drove 210 miles, how many miles was left for Mom to drive? *Ans:* 230 miles.

30. Two pumps sucked 170 gallons of water. If the first pumped 60 gallons, how much did the second do? *Ans:* 110 gallons.

31. A movie theater sold 310 tickets for 2 shows. If they sold 150 tickets for the first show, how many tickets were sold for the second? *Ans:* 160 tickets.

32. For a high school dance 410 students showed up. If there were 390 girls, how many boys came? *Ans:* 20 boys.

33. A collector divided 340 stamps between two albums. She put 150 stamps in one album, how many in the other? *Ans:* 190 stamps.

34. A grower planted 380 apple trees. By the end of the season 190 trees didn't make it. How many grew well? *Ans:* 190 trees.

35. Over the summer, 330 mosquitoes landed on Tony. He slapped 180. How many did he miss? *Ans:* 150 mosquitoes.

36. Mrs. Ruth had 260 baseball cards in her collection and sold 190 cards at the show. How many cards does she still have? *Ans:* 70 cards.

37. Two schools in town received 240 new desks. If one school took 70 desks, how many desks did the other school get? *Ans:* 170 desks.

38. The school bought 260 textbooks for the 1st and 2nd graders. If the 2nd graders received 140 books, how many books did they buy for the 1st graders? *Ans:* 120 books.

39. Two bricklayers laid 330 bricks. If the first laid 180 bricks, how many did the second bricklayer lay? *Ans:* 150 bricks.

40. One kid lied that he can lift 130 pounds, and the other kid lied that he can lift 190. How many pounds did both lie they can left? *Ans:* 320 pounds.

41. A store received 800 oz of olive oil and sold 190 oz to a restaurant. How many ounces of oil were left? *Ans:* 610 oz.

42. A carpenter bought 500 feet of baseboard and used 120 feet. How many feet of baseboard are left? *Ans:* 380 feet.

43. Coming back from her vacation, a busy senator found 600 unsigned papers on the desk. She signed 260 papers that day. How many more papers were left to sign? *Ans:* 340 papers.

44. The Central Park in New York City is 800 acres.
    a) If 150 acres are covered with water, how many is the land?
    Ans: 650 acres.
    b) If 240 acres are lawns, how many are not? *Ans:* 560 acres.
    c) If 130 acres are woodland, how many are not? *Ans:* 670 acres.
    Find Central Park on the map; it's in the middle of Manhattan
    Island in New York City. It has 26,000 trees, including 1,700
    American Elms, over 9,000 benches, and 150 drinking
    fountains. Every year 25 million people visit Central Park.

45. A building has 400 apartments; 170 of the apartments are
    studios, the rest are one-bedrooms. How many one-bedrooms
    are there? *Ans:* 230 apartments.

46. There are 300 channels on our television; 270 channels have
    boring and silly shows, the rest of the channels are not much
    better. How many "not much better" channels are there?
    Ans: 30 channels.

47. It takes 700 bags of gravel to pave the road. If 240 bags were
    delivered already, how many more bags do we need?
    Ans: 460 bags.

48. A picture with the frame costs $690. The frame alone costs
    $610. What's the price of the picture? *Ans:* $80. It happens.

49. At the factory, 170 computers out of 900 didn't pass quality
    control. How many computers passed? *Ans:* 730 computers.

50. For two buildings, a construction crew used 800 tons of
    cement. If they used 540 tons for one, how many tons of
    cement did they use for the other? *Ans:* 360 tons.

51. A mill received 520 sacks of wheat and 190 sacks of rye. How
    many sacks of both did they receive? *Ans:* 710 sacks.

ADDITION OF DOUBLE-DIGIT NUMBERS WITH SUM OVER 100

There are two ways to add double-digit numbers.

a) We separate tens and ones in both numbers, add tens to tens and one to ones and then add sums together.

b) The other way is first separate tens and ones of one number only. Next, first, add the tens and then ones to the other number.

Problem: 76 + 37 = ?

Solution 1: 76 is equal to 70 + 6; and 37 is equal 30 + 7.
Then, 70 + 30 = 100 and 6 + 7 = 13.
Together, 100 + 13 equals 113.

76 + 37 = 113

Solution 2: 37 is equal 30 + 7. Then 76 + 30 = 106,
and then 106 + 7 = 113.

EXERCISE I

1. What's the largest one-digit number? **Ans:** 9

2. What's the smallest two-digit number? **Ans:** 10

3. What's the largest two-digit number? *Ans:* 99

4. What's the smallest three-digit number? *Ans:* 100

5. What's the largest three-digit number? *Ans:* 999

6. What's the smallest one-digit number? *Ans:* 0 (don't forget 0)

7. What's the largest four-digit number? *Ans:* 9999

8. What's the smallest four-digit number? *Ans:* 1000

9. What's the largest five-digit number? *Ans:* 99999

10. How many digits are on one hand? *Ans:* 5. Remember, in Latin digit means a finger, and the thumb counts as one digit.

▶ *A trick: Adding numbers that end in 9:* When one of the numbers ends with 9, it's easier to add 1 to that number to convert it to a number that ends with 0, subtract 1 from the other number, and then add them together.

Problem: 19 + 25 = ?
Solution: First, add 1 to 19 to make 20.
Next, subtract 1 from 25 to make 24.
Then, 20 + 24 is equal to 44.

19 + 25 = 44

Problem: 59 + 39 = ?
Solution: First, 59 + 1 = 60. Then, 39 - 1 = 38. Then, 60 + 38 = 98.

59 + 39 = ? *Ans:* 98	64 + 19 = ? *Ans:* 83	28 + 99 = ? *Ans:* 127
39 + 32 = ? *Ans:* 71	19 + 19 = ? *Ans:* 38	89 + 20 = ? *Ans:* 109
36 + 29 = ? *Ans:* 65	29 + 29 = ? *Ans:* 58	67 + 59 = ? *Ans:* 126
49 + 24 = ? *Ans:* 73	99 + 54 = ? *Ans:* 153	25 + 29 = ? *Ans:* 54
25 + 29 = ? *Ans:* 54	69 + 28 = ? *Ans:* 97	25 + 29 = ? *Ans:* 54

EXERCISE II

73 + 30 = ? *Ans:* 103	36 + 90 = ? *Ans:* 126	80 + 58 = ? *Ans:* 138
55 + 60 = ? *Ans:* 115	66 + 70 = ? *Ans:* 136	70 + 87 = ? *Ans:* 157
50 + 70 = ? *Ans:* 120	60 + 90 = ? *Ans:* 150	70 + 47 = ? *Ans:* 117
56 + 70 = ? *Ans:* 126	60 + 94 = ? *Ans:* 154	33 + 80 = ? *Ans:* 113

60 + 82 = ? *Ans:* 142	75 + 70 = ? *Ans:* 145	36 + 90 = ? *Ans:* 126
40 + 94 = ? *Ans:* 134	69 + 90 = ? *Ans:* 159	66 + 72 = ? *Ans:* 138

EXERCISE III

60 + 43 = ? *Ans:* 103	69 + 90 = ? *Ans:* 159	67 + 33 = ? *Ans:* 100
70 + 64 = ? *Ans:* 134	60 + 71 = ? *Ans:* 131	55 + 53 = ? *Ans:* 108
60 + 49 = ? *Ans:* 109	83 + 80 = ? *Ans:* 163	72 + 52 = ? *Ans:* 124
35 + 80 = ? *Ans:* 115	88 + 50 = ? *Ans:* 138	71 + 38 = ? *Ans:* 109
94 + 50 = ? *Ans:* 144	66 + 60 = ? *Ans:* 126	94 + 44 = ? *Ans:* 138
78 + 60 = ? *Ans:* 138	60 + 87 = ? *Ans:* 147	94 + 36 = ? *Ans:* 130

EXERCISE IV

75 + 75 = ? *Ans:* 150	73 + 68 = ? *Ans:* 141	87 + 39 = ? *Ans:* 126
76 + 76 = ? *Ans:* 152	46 + 76 = ? *Ans:* 122	58 + 78 = ? *Ans:* 136
84 + 82 = ? *Ans:* 166	55 + 58 = ? *Ans:* 111	66 + 66 = ? *Ans:* 132
88 + 82 = ? *Ans:* 170	86 + 58 = ? *Ans:* 144	72 + 79 = ? *Ans:* 151
83 + 55 = ? *Ans:* 138	95 + 74 = ? *Ans:* 169	46 + 74 = ? *Ans:* 120
65 + 53 = ? *Ans:* 118	76 + 59 = ? *Ans:* 135	47+ 78 = ? *Ans:* 125

WORD PROBLEMS

1. Greg can lift 54 pounds with each hand. How many pounds can he lift with both hands? *Ans:* 108 pounds.

2. Last year the hospital hired 55 nurses. This year they hired 55 more. How many nurses did the hospital hire in two years? *Ans:* 110 nurses.

3. In each issue of a teen magazine there are 51 ads.
   a) How many ads are in two issues? *Ans:* 102 ads.
   b) How many ads are in three issues? *Ans:* 153 ads.

4. An adult ticket costs $72 and child's ticket is $31. How much do both tickets cost? *Ans:* $103.

5. If one twin is 52 inches tall, what's the height of two twins, one on top of the other? *Ans:* 104 inches.

6. There are 67 words on one page and 42 on the other. How many words are on both pages? *Ans:* 109 words.

7. A boat's gas tank takes 84 gallons of gasoline. A truck's tank takes 24 gallons. How many gallons do both take? *Ans:* 108 gallons.

8. One candy box has 56 candies. How many candies are in two boxes? *Ans:* 112 candies.

9. One windmill turbine makes 83 watts of electricity, another makes 30 watts. How many watts do both generate? *Ans:* 113 watts. We use watt as the measure of electric power.

10. The store received 62 pairs of men shoes and 41 pairs of women shoes. How many shoes did the store receive? *Ans:* 206 shoes. Don't get tricked. The store received 62 + 41 = 103 pairs of shoes. Each pair is made of two shoes. Then, 103 + 103 = 206 shoes.

11. Bees from two different beehives are sharing a rose bush. One beehive is 55 feet away from the rose bush. The other is 65 feet away in the opposite site. How many feet are between two beehives? *Ans:* 120 feet. If you miss the rose bush you can visit the neighbors.

12. One beehive produced 74 ounces of honey, the other made 83. How much honey did both make? *Ans:* 157 ounces. Sweet deal!

13. One package weighs 37 ounces, the other weighs 80 ounces more. How much does the second package weigh? *Ans:* 117 ounces.

14. There are 59 short and 49 long stories in the book. How many stories are in the book? *Ans:* 108 stories.

15. A restaurant served 55 cups of tea and 64 cups of coffee. How many cups did they serve? *Ans:* 119 cups.

16. One building has 74 offices, and the building next door has 55. How many offices are in both buildings? *Ans:* 129 offices.

17. I planted 137 carrot seeds and 63 tomato seeds. How many seeds did I plant? *Ans:* 200 seeds.

18. Paolo counted 60 small towns and 75 villages on the map. How many of both did he count? *Ans:* 135 small town and villages.

19. There are 64 miles from Paolo's home to the ocean. How long is the round-trip? *Ans:* 128 miles.

20. A desk chair costs $57. How much will two chairs cost? *Ans:* $114.

21. Mrs. Crust removed 157 apple seeds and 43 orange seeds from her salad. How many seeds did she take out? *Ans:* 200 seeds.

22. There are 78 men and 27 women on a SWAT team. How many people are on the team? *Ans:* 105 members.

23. A nurse gave 88 shots and 31 finger sticks. How many times did the nurse poke the kids? *Ans:* 119 times.

24. On the beach Louis found 34 shells and 77 pebbles. How many of both did Louis find? *Ans:* 111 shells and pebbles.

25. Grandma Wendy's 100th birthday is January 1st. She received 143 Happy Birthday cards and 77 Happy New Year cards. How many cards did she get? *Ans:* 220 cards.

26. If one chessboard has 64 squares, how many squares are on two boards? *Ans:* 128 squares.

27. In a big town, 76 stations sell only gasoline and 34 stations sell gasoline and diesel fuel. How many stations are in town? *Ans:* 110 stations.

28. The town has 67 police officers and 46 deputy sheriffs. How many peace officers are in town? *Ans:* 113 officers.

29. At the City College, the art department has 84 professors and the science department has 55. How many professors are in both departments? *Ans:* 139 professors.

30. At the College there are 73 students on the football team and 57 students on the basketball team. How many students are on both teams? *Ans:* 130 students.

31. One water pump pumps 89 gallons per hour, the other 59 gallons per hour. How many gallons do they pump together? *Ans:* 148 gallons.

32. A raccoon found 69 pears under one tree and 42 pears under the other. How many pears did it find? *Ans:* 111 pears.

33. At a science fair, Blair's bubble machine made 56 bubbles. Cody's machine made 55. How many bubbles did both make? *Ans:* 111 bubbles.

34. Pickle Monster ate 38 pickled mushrooms after eating 83 pickled tomatoes. How many pickled vegetables did it eat? *Ans:* 121 pickles.

35. There are 67 cats and 86 dogs in our town. How many cats and dogs are there? *Ans:* You probably shouldn't add dogs to cats or cats to dogs, but if you do, the answer is 153 cats and dogs.

36. A newspaper received 72 subscription requests by mail and 49 by fax. How many requests did they receive? *Ans:* 121 subscriptions.

37. One constellation has 93 stars, the other 58. How many stars are in both constellations? *Ans:* 151 stars.

38. A famous mathematician lived 38 years in XIX century and 37 years in the XX century. How many years did he live? *Ans:* 75 years.

39. A brave cat caught 58 mice in one week. How many mice could it catch in two weeks? *Ans:* 116 mice.

40. A brave mouse ran away from 62 cats in one week. From how many cats could it run away from in two weeks? *Ans:* 124 cats.

41. If Mrs. Cockatiel puts 77 parrots in one cage and 46 parrots in the other, how many parrots will be in both cages? *Ans:* 123 parrots.

42. Mrs. Cockatiel took 49 parakeets and put them in a cage with 52 parakeets. How many birds are in the cage now? *Ans:* 101 birds.
   Do you know the difference between parrots and parakeets? The word parakeet means small parrot. The birds look very similar; parakeets, also known as budgies, are smaller but have longer tails.

43. If Inga caught 67 snowflakes and Nielsen caught 74, how many snowflakes did they catch together? *Ans:* 141 snowflakes.

44. If a canoe is 84 inches long and a kayak is 86 inches long, what's the length of both put in line? *Ans:* 169 inches.
    Do you know the difference between canoe and kayak? Kayaks were invented by the Eskimos and are narrow. Canoes were invented by Native North Americans and are broader. You paddle kayak sitting down with a double bladed paddle. A canoe is paddled kneeling or sitting using a single bladed paddle.

45. If a translator can translate 57 words per minute, how many words can she translate in two minutes? *Ans:* 114 words.

46. If a baby can rip a paper into 59 small pieces in 1 minute, how many pieces can the baby make in 2 minutes? *Ans:* 118 pieces.

47. If a clown falls down 64 times in one hour, how many times will he fall in two hours? *Ans:* 128 times.

48. If a team of electricians replaced 75 switches in one day, how many switches will they replace in two days? *Ans:* 150 switches.

49. If a drummer hits a drum 88 times a minute, how many times does he hit in two minutes? *Ans:* 176 times.

50. If a troublemaker did 66 graffiti in one day before being arrested by the police, how many graffiti he could have done in two days? *Ans:* 132 graffiti.

SUBTRACTING DOUBLE-DIGIT NUMBERS FROM A TRIPLE-DIGIT NUMBERS

We break the number we subtract into tens and ones. Then, we take them from minuend (remember, that's the number you are taking from) one at a time.

Problem: 184 - 52 = ?
Solution: 52 is equal to 50 and 2. Next, 184 - 50 = 234, and then, 134 - 2 = 132.

184 - 52 = 132

Problem: 118 - 33 = ?
Solution: 33 is equal to 30 plus 3. Next, 118 - 30 = 88; then, 88 - 3 = 85

118 - 33 = 85

Problem: 224 - 67 = ?
Solution: 67 is equal to 60 + 7. Next, 224 - 60 = 164, then, 164 - 7 = 157

224 - 67 = 157

Another easy way to do this is to think, I can subtract 224 - 64 quickly, and that is 160. Now have to take 3 more, so the answer is 157.

Always start from the left side when doing addition or subtraction in your head and try to simplify the problem. Do not try to do it as you would when you do it on paper, starting from right side. That is too hard to do in your head.

EXERCISE I

200 - 30 = ? *Ans:* 170	329 - 50 = ? *Ans:* 279	529 - 80 = ? *Ans:* 449
245 - 40 = ? *Ans:* 205	616 - 90 = ? *Ans:* 526	209 - 90 = ? *Ans:* 119
400 - 70 = ? *Ans:* 330	300 - 80 = ? *Ans:* 220	467 - 80 = ? *Ans:* 387
408 - 70 = ? *Ans:* 338	325 - 80 = ? *Ans:* 245	546 - 60 = ? *Ans:* 486
200 - 60 = ? *Ans:* 140	432 - 60 = ? *Ans:* 372	612 - 90 = ? *Ans:* 528
211 - 60 = ? *Ans:* 151	612 - 30 = ? *Ans:* 582	616 - 90 = ? *Ans:* 526

EXERCISE II

1. How much do you add to 10 to get 30? *Ans:* 20

2. How much do you add to 12 to get 20? *Ans:* 8

3. How much do you add to 45 to get 100? *Ans:* 55

4. How much do you add to 38 to get 100? *Ans:* 62

5. How much do you add to 47 to get 100? *Ans:* 53

6. How much do you add to 12 to get 100? *Ans:* 88

7. How much do you add to 17 to get 100? *Ans:* 83

8. How much do you add to 56 to get 100? *Ans:* 44

9. How much do you add to 33 to get 100? *Ans:* 67

10. How much do you add to 65 to get 100? *Ans:* 35

11. How much do you add to 59 to get 100? *Ans:* 41

12. How much do you add to 54 to get 100? *Ans:* 46

13. How much do you add to 57 to get 100? *Ans:* 43

14. How much do you add to 85 to get 100? *Ans:* 15

15. How much do you add to 16 to get 100? *Ans:* 84

16. How much do you add to 75 to get 100? *Ans:* 25

17. How much do you add to 21 to get 100? *Ans:* 79

EXERCISE III

Problem: What do you take away from 30 to get the number equal to 9 + 5?

Solution: This is a two-step problem. First, we find the sum of 9 and 5, which is 14. Then, we take 14 from 30, 30 - 14 = 16. Therefore, the answer is the number we need to take from 30 to get the sum of 9 and 6 is 16.

Problem: What do you take away from 100 to get number equal to sum of 25 and 5? *Solution:* The sum of 25 and 5 is 30. Then, 100 - 30 = 70. *Ans:* 70

1. What do you take away from 100 to get the number equal to the sum of 65 and 35? *Ans:* 0

2. What do you subtract from 100 to get the number equal to the sum of 7 and 22? *Ans:* 71

3. What do you subtract from 100 to get the number equal to the sum of 12 and 19? *Ans:* 69

4. What do you subtract from 100 to get the number equal to the sum of 23 and 34? *Ans:* 43

5. What do you subtract from 100 to get the number equal to the sum of 17 and 22? *Ans:* 61

6. What do you subtract from 100 to get the number equal to the sum of 16 and 28? *Ans:* 56

7. What do you subtract from 100 to get the number equal to the sum of 50 and 15? *Ans:* 35

8. What do you subtract from 100 to get the number equal to the sum of 42 and 35? *Ans:* 23

EXERCISE IV

100 - 64 = ? *Ans:* 36	100 - 37 = ? *Ans:* 63	200 - 25 = ? *Ans:* 175
100 - 43 = ? *Ans:* 57	100 - 59 = ? *Ans:* 41	200 - 75 = ? *Ans:* 125
100 - 22 = ? *Ans:* 78	100 - 25 = ? *Ans:* 75	110 - 61 = ? *Ans:* 49
100 - 19 = ? *Ans:* 81	110 - 25 = ? *Ans:* 85	100 - 38 = ? *Ans:* 62
100 - 86 = ? *Ans:* 14	120 - 25 = ? *Ans:* 95	120 - 68 = ? *Ans:* 52

106 - 54 = ? *Ans:* 52	112 - 54 = ? *Ans:* 58	146 - 81 = ? *Ans:* 65
158 - 44 = ? *Ans:* 114	142 - 75 = ? *Ans:* 67	260 - 99 = ? *Ans:* 161
134 - 45 = ? *Ans:* 89	120 - 59 = ? *Ans:* 61	145 - 46 = ? *Ans:* 99

WORD PROBLEMS

1. A truck delivered 103 watermelons to a store, 13 watermelons broke on the way. How many watermelons are good to be sold? *Ans:* 90 watermelons.

2. There were 109 new vocabulary words on the test. Milo made 16 mistakes. How many words were correct? *Ans:* 93 words.

3. A repair shop fixed 109 computers. There were 24 laptops and the rest were desktops. How many desktops did they repair? *Ans:* 85 desktops.

4. Out of 109 computers the shop repaired, 36 were Macs and the rest PC computers. How many PC's did they fix? *Ans:* 73 PC's.

5. It takes 106 days to cross the United States on foot from coast to coast. It takes 11 days on the bicycle. How many more days does it take to walk than bike?
   Ans: 95 days. Did you know that the distance from coast to coast is about 3,000 miles?

6. There were 114 gallons of gas in the truck's tank. It took 41 gallons to get to the destination. How many gallons of gas are in the tank now? *Ans:* 73 gallons.

7. In two days a scientist read 105 scientific papers. How many papers did she read the first day if on the second day she read 29? *Ans:* 76 papers.

8. A turtle weighs 103 pounds but without the shell it weighs only 24 pounds. How much does the shell weigh? *Ans:* 79 pounds. How do you talk the turtle into taking off its shell?

9. Mama whale and baby whale weigh 130 tons. What's mama's weight if the baby is 33 tons? *Ans:* 97 tons.

10. The weight of the bone on a 120 ounces steak is 32 ounces. What's the weight of the steak without the bone?
    Ans: 88 ounces.

11. Every week Mrs. Fiesta sleeps 56 out of 168 hours. How many hours is she awake? *Ans:* 112 hours.

12. If out of 112 wake hours Mrs. Fiesta spends 40 at work, how many hours are left for everything else during the week? *Ans:* 72 hours.

13. Out of 72 hours, Mom spent 33 with her children and the rest for housekeeping. How many hours were left for housekeeping? *Ans:* 39 hours.

14. A food bank received 132 cans in one day. 55 were soups and the rest were veggies. How many veggie cans did the bank receive? *Ans:* 77 cans.

15. After spilling tomato sauce, Homer used 27 paper towel sheets of to wipe the floor. If the new roll had 110 pieces, how many were left? *Ans:* 83 pieces, more than enough for 3 more accidental spills.

16. Silvia bought a salami, 111 centimeters long. On the way home and hungry she bit off 13 centimeters (cord and all).
a) How much salami did she bring home? *Ans:* 98 centimeters.
b) At home she cut another 19 centimeters from what was left. How much salami remained? *Ans:* 79 centimeters.
c) If she cuts another 22 centimeters from what's left, how much salami would be then? *Ans:* 57 centimeters of salami.

17. A pine cone had 123 pine nuts before a squirrel ate 47. How many nuts were left in the cone? *Ans:* 76 nuts.

18. The family picture was on the wall for 102 days before we noticed it was upside down. If Timmy confessed playing the trick 45 days after the picture was hanged, how many days was it hanging upside down? *Ans:* 57 days.

19. One day, a speedy sparrow caught 125 flies. The next day it caught 38 more flies. How many flies did it catch the second day? *Ans:* 163 flies. Not bad.

20. The width of the carpet is 48 inches less than its length. If the carpet is 124 inches long, what's the width? *Ans:* 76 inches.

21. Two magazines together have 128 pages. If the first is 64 pages long, how many pages are in the second magazine? ***Ans:*** Also 64 pages.

22. A brave mongoose bit off 47 inches off a 120 inch long deadly cobra. How much is left of the snake? ***Ans:*** 73 inches. Mongooses are about 4 feet long or less. They feed on insects, snakes, and rats. King cobras, the world's longest poisonous snake, can reach 18 feet in length.

23. Out of 114 coins in Conrad's piggy bank, 67 are quarters and the rest are nickels and dimes. How many nickels and dimes are in the bank? ***Ans:*** 47 nickels and dimes.

24. Chen put 143 green peas in the bird feeder. The birds ate 77 peas and dropped the rest on the ground. How many peas are on the ground? ***Ans:*** 66 peas.

25. A race car drove at 103 miles per hour. If the wind was blowing at 29 miles per hour, what was the car's speed without the wind? ***Ans:*** 74 mile per hour.

26. A 111 page book has 58 pages with illustrations. How many pages only have text ? ***Ans:*** 53 pages.

27. A mail carrier started the day with 146 letters and by noon still had 85 letters in his bag. How many letters did he deliver before noon? ***Ans:*** 61 letters.

28. Another mail carrier has 163 pieces of mail. There are 87 letters and the rest are magazines. How many magazines does she have? ***Ans:*** 76 magazines.

29. A delivery truck with 158 lizards on board dropped 78 lizards at a pet store. How many lizards stayed on? ***Ans:*** 80 lizards.

30. A fireman, a paramedic and a policeman together bought 122 doughnuts. If the fireman bought 59 doughnuts and the paramedic bought 63, how many doughnuts did the policeman buy? ***Ans:*** 0 (none). Who said that police officers like doughnuts?

31. There were 110 skaters at the skate rink.
    a) If 45 skaters wore hats, how many didn't? *Ans:* 65 skaters.
    b) If there were 53 girl skaters, how many boys were in the rink? *Ans:* 57 boys.
    c) If 64 skaters wore gloves and the rest wore mittens, how many skaters wore mittens? *Ans:* 46 skaters.
    d) If only 38 skaters didn't fall down, how many fell? *Ans:* 72 skaters.

32. Liana tried to cut a 134 inch rope in half, but instead cut off 65 inches. How long was the remaining piece? *Ans:* 69 inches.

33. From a pile of 120 planks, Reese took 57 to make a birdhouse. Does she have enough to make a second house? *Ans:* Yes, because there are 63 planks left.

34. A gardener planted 130 tulip bulbs and watched 75 of them grow. How many haven't come up yet? *Ans:* 55 tulips. Tulips like cold and if planted in warm climate need to be refrigerated for several weeks. When planting tulips it's important to dig a deep hole. Tulips planted deeper have thicker stems and don't fall over easily.

35. A library has 153 DVDs and 77 CDs. How many more DVDs are in the library? *Ans:* 76 more DVDs.

36. Out of 153 videos, 85 are fun movies and the rest are educational. How many educational videos are in the library? *Ans:* 68 educational videos.

37. 79 butterflies came out of 131 cocoons. How many butterflies are still in cocoons? *Ans:* 52 butterflies.

38. There were 150 silkworms in a jar and Jeff took out 83. How many worms are in the jar now? *Ans:* 67 worms. Silkworms are a special type of caterpillar. They were discovered and farmed in China more than 5,000 years ago.

39. In the morning there were 134 ice cream trucks on the parking lot. By midday there were only 69. How many trucks left early? *Ans:* 65 trucks.

40. Ulla hoped to read 120 books over the summer. By August 1st she read 77. How many more books are left for her to read? *Ans:* 43 books.

41. Mrs. Lopez is 103 years old. She lived in Cuba for 47 years before moving to Florida. How long has she been living in Florida? *Ans:* 56 years.
The oldest person was probably Jeanne Louise Calment who lived 122 years and 164 days.

42. The road through the mountains is 164 miles long. 89 of those miles are uphill. How long is the downhill part of the road? *Ans:* 75 miles.

43. Out of 187 fans that came to the volleyball game, 99 were rooting for the home team. How many cheered for the guest team? *Ans:* 88 fans.

44. Michael crossed the river in 87 strokes and crossed back in 105 strokes. How many strokes did he use both ways? *Ans:* 192 strokes.

45. Gigi bought 150 carrots and fed 75 to her rabbits. How many are left? *Ans:* 75 carrots.

46. In one minute a rabbit runs 115 feet and a tortoise runs 87 feet less. How many feet does a tortoise run in one minute? *Ans:* 28 feet.

47. By mistake, Tamir put 98 guppies in a tank with 15 sharks.
a) How many fish were in the tank total? *Ans:* 113 fish.
b) How many fish were eaten, if there are only 37 fish left in the tank? *Ans:* 76 fish, all guppies, because sharks eat guppies but guppies don't eat sharks.

48. If there are 158 bees in two beehives and one beehive has 79 bees, how many bees are in the second? *Ans:* 79 bees.

49. On a field trip to a ceramic shop, the children made 143 clay figurines. The kiln oven can fit in only 87 items at a time. How many figurines didn't make the first firing? *Ans:* 56 figurines.

50. A movie director invited 183 actors for her new movie. There were 89 adults, and the rest were kids. How many children actors were in the movie? *Ans:* 94 children actors.

ADDITION AND SUBTRACTION OF DOUBLE AND TRIPLE-DIGIT NUMBERS

To *add* triple-digit numbers in your head, break one of the numbers into hundreds, tens, and ones. Then add them one at a time to the other number. Don't try to do it from the right like you do on paper! That will be too slow.

We break the number we subtract into hundreds, tens, and ones. Then, we take them out one at a time.

Problem: 244 + 135 = ?

Solution: Start at the left. In adding numbers in the head, its always better to start at the left side of the number. Add 2 and 1 and we see that the answer will be about 300. Now look at the last two numbers, they are 44 and 35 which are easy to add to 79, so the total answer is 379.

Problem: 176 + 204 = ?

Solution: Start at the left. Add 1 and 2 and we see that the answer will be about 300. Now look at the last two numbers, they are 76 and 4 which are easy to add to 80, so the total answer is 380.

Problem: 281 - 133 = ?

Solution: Start at the left. Subtract 1 from 2 and we see that the answer will be about 100. Now look at the last two numbers, they are 81 and 33 the difference of which is by the same method, 44, so the total answer is 144.

Problem: 460 - 352 = ?

Solution: Start at the left. Subtract 3 from 5 and we see that the answer will be about 100. Now look at the last two numbers, they are 60 and 52 the difference of which is by the same method, 8, so the total answer is 108.

EXERCISE I

Do these by working from the left side.

65 + 35 = ? *Ans:* 100	64 + 54 = ? *Ans:* 118	41 - 26 = ? *Ans:* 15
52 + 54 = ? *Ans:* 106	44 + 95 = ? *Ans:* 139	83 - 36 = ? *Ans:* 47
58 + 72 = ? *Ans:* 130	65 + 43 = ? *Ans:* 108	164 - 29 = ? *Ans:* 135
83 + 84 = ? *Ans:* 167	80 - 55 = ? *Ans:* 25	260 - 41 = ? *Ans:* 219
61 + 96 = ? *Ans:* 157	77 - 38 = ? *Ans:* 39	151 - 49 = ? *Ans:* 102
52 + 84 = ? *Ans:* 136	44 - 18 = ? *Ans:* 26	366 - 38 = ? *Ans:* 328
75 + 75 = ? *Ans:* 150	73 - 40 = ? *Ans:* 33	440 - 250 = ? *Ans:* 190

▶ *A trick: Subtracting numbers ending with 9 or 1.* When subtracting a number ending with 1 from another number, take away 1 with both numbers before doing subtraction, it might be easier this way.

Problem: 78 - 31 = ?

Solution: First, take away 1 from both and get 77 and 30. Then, 77 - 30 = 47

78 - 31 = 47

EXERCISE II

37 - 11 = ? *Ans:* 26	71 - 41 = ? *Ans:* 30	70 - 51 = ? *Ans:* 19
48 - 41 = ? *Ans:* 7	87 - 41 = ? *Ans:* 46	44 - 21 = ? *Ans:* 23
33 - 21 = ? *Ans:* 12	59 - 31 = ? *Ans:* 28	69 - 51 = ? *Ans:* 18
53 - 21 = ? *Ans:* 32	94 - 61 = ? *Ans:* 33	86 - 51 = ? *Ans:* 35

When subtracting a number ending with 9 from another number, add 1 to both numbers and then subtract.

Problem: 63 - 29 = ?
Solution: First, add 1 to both and get 64 and 30. Then, 64 - 30 = 34.

63 - 29 = 34

57 - 9 = ? *Ans:* 48	55 - 39 = ? *Ans:*16	67 - 31 = ? *Ans:* 36
14 - 9 = ? *Ans:* 5	62 - 39 = ? *Ans:* 23	74 - 41 = ? *Ans:* 33
36 - 19 = ? *Ans:* 17	79 - 29 = ? *Ans:* 50	95 - 51 = ? *Ans:* 44
48 - 39 = ? *Ans:* 9	87 - 69 = ? *Ans:* 18	106 - 61 = ? *Ans:* 45
54 - 29 = ? *Ans:* 25	61 - 49 = ? *Ans:* 12	34 - 19 = ? *Ans:* 15
48 - 19 = ? *Ans:* 29	93 - 49 = ? *Ans:* 44	24 - 19 = ? *Ans:* 5

EXERCISE III

215 + 105 = ? *Ans:* 320	218 - 107 = ? *Ans:* 111
126 + 310 = ? *Ans:* 436	245 - 124 = ? *Ans:* 121
247 + 102 = ? *Ans:* 349	356 - 243 = ? *Ans:* 113
204 + 134 = ? *Ans:* 338	468 - 241 = ? *Ans:* 227
170 +106 = ? *Ans:* 276	345 - 106 = ? *Ans:* 239
307 + 209 = ? *Ans:* 516	447 - 50 = ? *Ans:* 397
108 + 180 = ? *Ans:* 288	453 - 224 = ? *Ans:* 229
244 + 136 = ? *Ans:* 380	450 - 245 = ? *Ans:* 205
143 + 255 = ? *Ans:* 398	990 + 450 = ? *Ans:* 540
111 + 803 = ? *Ans:* 914	423 - 214 = ? *Ans:* 209
424 - 324 = ? *Ans:* 100	

EXERCISE IV

Adding the equal double-digit numbers:

The easy way to add several double digit numbers is to add all the tens first, then add all the ones, and then add tens and ones together.

Example: 13 + 13 + 13 will equal to adding first 10 + 10 + 10 = 30. Next, 3 + 3 + 3 = 9 (you can also multiply, 3 × 3 =9). Then, 30 + 9 = 39.

Problem:	34 + 34 + 34 = ?
Solution:	30 + 30 + 30 = 90. Next, 4 + 4 + 4 = 12 (or 4 × 3 = 12).
	Then, 90 + 12 = 102
	34 + 34 + 34 = 102

Read first three numbers to the child.

16 + 16 + 16 = 48	38 + 38 + 38 = 114
25 + 25 + 25 = 75	47 + 47 + 47 = 141
32 + 32 + 32 = 96	55 + 55 + 55 = 165
35 + 35 + 35 = 105	17 + 17 + 17 = 51
41 + 41 + 41 = 123	65 + 65 + 65 = 195
28 + 28 + 28 = 84	39 + 39 + 39 = 127
53 + 53 + 53 = 159	15 + 15 + 15 = 60
29 + 29 + 29 = 87	12 + 12 + 12 = 36

WORD PROBLEMS

1. This week, 122 new books were donated to the school library. If there were 101 books to start, how many books are in the library now? *Ans:* 223 books. Do you donate books to your library?

2. The library staff put plastic covers on 116 regular books and 92 oversized books. How many books did they cover?
    Ans: 208 books.

3. Miranda knew 80 jokes and learned 39 new ones. How many jokes does she know now? *Ans:* 119 jokes.

4. A princess has 291 handkerchiefs; 70 hankies are mono-grammed. How many are not? *Ans:* 221 handkerchiefs. Monogram is a design of person's one or more initials.

5. One computer game uses 151 megabytes of memory, another uses 110. How much memory do both use?
    Ans: 261 megabytes.

6. There are 205 nonfiction books and 130 reference books in the library. How many books are there? *Ans:* 335 books. Reference book might be a dictionary or an encyclopedia where you can check for reliable facts.

7. There are 260 hazelnuts in two bags. How many nuts are in the first bag if the second has 105 nuts? *Ans:* 155 nuts.

8. There are 170 hens and 119 roosters on the farm. How many birds live there? *Ans:* 289 birds.

9. A motorcyclist drove 153 miles the first day and 50 miles less the second day. How many miles did she drive in two days? *Ans:* 256 miles (153 - 50 = 103, then 153 + 103 = 256).

10. There are 170 lbs of carrots in one bag and 99 lbs less in the other. How many carrots are in the second bag? *Ans:* 71 lbs.

11. There are 640 pages in a two-volume Shakespeare play collection. If the first volume has 330 pages, how many pages are in the second volume? *Ans:* 310 pages.
    Shakespeare wrote 37 known plays and also many poems and sonnets.

12. For the river cruise, 672 people boarded two boats. If there were 250 passengers on one boat, how many were on the other? *Ans:* 422 passengers.
    Do you know the difference between a ship and a boat?

13. One truck carries 344 solar panels, the other truck has 160 more panels. How many panels are on the second truck? *Ans:* 504 panels.

14. At a campsite there are 170 scorpions and 137 poisonous tarantulas. How many critters are there? *Ans:* 307 critters.

15. The same camp had 59 rattle snakes. How many nasty things were there then? *Ans:* 366 nasty things.

16. One rattlesnake is 170 centimeters long, another is 66 centimeters shorter. How long is the second snake?
    Ans: 104 centimeters.

17. Lily and May entered a high speed reading competition. In one hour, Lily read 117 pages and May read 120. How many pages did both read? *Ans:* 237 pages.

18. There are 530 feet across the lake but only 311 feet to the rock in the middle. How many feet are from the rock to the other side? *Ans:* 219 feet.

19. It takes 50 minutes to come down from the mountain and 183 minutes to get to the top.
    a) How long does it take to get to the top and then down?
    Ans: 233 minutes.
    b) How much longer does it take to the top than getting down?
    Ans: 133 minutes longer.

20. A poet bought 275 sheets of paper, then wrote on and tore 140 of them. How many more pages are left for writing and tearing? *Ans:* 135 pages.

21. In one year Simon shed 117 tears from his right eye and 80 tears from his left. How many fell from both eyes? *Ans:* 197 tear drops. Simon is only two.

22. At a meeting, one professor used 170 fancy words and another 52 less words. How many fancy words did both use?
    Ans: 288 words (170 - 52 = 118, then 170 + 118 = 288). Fancy words like derivative or integral.

23. A shepherd cut 280 pounds of wool from the white sheep and 121 pounds from the black sheep. How many pounds did he cut altogether? *Ans:* 401 pounds.

24. Using spinning wheel, a maiden spun 105 feet of yarn in one day. The next day she spun 20 yards less. How many feet of yarn did she spin in two days?
    Ans: 190 feet (105 - 20 = 85, then 105 + 85 = 190). Spinning is turning natural fibers, such as wool or cotton, into a thread. The spinning wheel, a clever device, was probably invented in India and came to Europe between the 13th and 14th centuries replacing spinning by hand.

25. One board is 263 inches long; the other is 110 inches shorter. How long is the second board? *Ans:* 153 inches.

26. One airplane flies at 340 miles per hour; the other is 124 mph slower. How fast is the second plane? *Ans:* 216 miles per hour. Some supersonic jets fly faster than sound which travels at over 760 miles per hour.

27. A well is 860 inches deep. The water in the well is 230 inches deep. What's the distance from the water surface to the top of the well? *Ans:* 630 inches.

28. From a 600 inch wire, an electrician cut 212 inches. How many inches were left in the wire? *Ans:* 388 inches.

29. A medieval manuscript has 350 pages; of these, 133 pages are unreadable. How many pages can be read? *Ans:* 217 pages. The word manuscript means "written by hand". Until 1439, when Johann Gutenberg invented printing press, all books in Europe were manuscripts.

30. If 449 pancakes were served with maple syrup and 550 pancakes with jam, how many pancakes were served? *Ans:* 999 pancakes.

31. An author typed 257 pages of a new mystery novel and has 150 more pages left to type. How long is the novel? *Ans:* 407 pages.

32. The willow in the park lived 99 years but the oak lived twice as long. How many years did the oak live?
Ans: 198 years (remember the rule of adding 1 to the first number and subtracting 1 from the other before adding them together).

33. Out of 550 restaurants in a city, 99 are vegetarian. How many non-vegetarian restaurants are in the city?
Ans: 451 restaurants (remember the rule of adding 1 to both numbers before subtracting).

34. A roofer used 700 shingles for the house and the garage. How many shingles did he use for the house if the garage roof needed 221 shingles? *Ans:* 479 shingles.

35. Two printers together printed 800 pages. How many pages did the second print, if the first turned out 335 pages?
Ans: 465 pages.

36. If a farm used 268 boxes of plums for making prunes and 220 boxes for making jam, how many boxes did they use?
Ans: 488 boxes.

37. Out of 400 quail eggs only 119 hatched. How many eggs didn't hatch? *Ans:* 281 eggs.

38. In Ariel's room, 148 CD's are on the shelf and 160 on the floor. How many CDS's are in Ariel's room? ***Ans:*** 308 CD's.

39. There are 321 French and 123 Italian passengers on a ferry. How many passengers are there? ***Ans:*** 444 passengers.

40. If a refrigerator's uses 236 watts and a desktop computer's 254 watts, what is the power of both? ***Ans:*** 490 watts. Did you know that some computers use more power than refrigerators?

41. One sumo wrestler weighs 345 pounds and another weighs 315 pounds.
a) How much do both weigh? ***Ans:*** 660 pounds.
b) By how much is the first wrestler heavier than the second? ***Ans:*** by 30 pounds.

42. There are 212 quarters in one coin-operated telescope and 217 in the other. How many quarters are in both? ***Ans:*** 429 coins.

43. Two trucks carried 480 pineapples. If the first truck had 202 pineapples, how many were in the second? ***Ans:*** 278 pineapples.

44. After 213 out of 312 cows crossed the river, how many are still on the other side? ***Ans:*** 99 cows.

45. Today, 271 students took the test but only 221 passed. How many failed the test? ***Ans:*** 50 students.

46. Out of 214 houses in the village, only 99 have tile roof. How many houses don't? ***Ans:*** 115 houses don't have tile roof.

47. Out of 322 mangoes, 99 are rotten. How many are still good? ***Ans:*** 223 mangoes.

48. I picked up 329 pinecones and then 199 more. How many pinecones did I pick up? ***Ans:*** 528 cones.

49. If a sombrero costs $17, what is the price of 4 sombreros? ***Ans:*** $68. A sombrero is a Mexican hat with wide brim, broad enough to cast a shadow over the head, neck and shoulders.

50. An office has 18 chairs.
a) How many chairs do 2 offices have? ***Ans:*** 36 chairs.
b) How many chairs do 3 offices have? ***Ans:*** 54 chairs.
c) How many chairs do 4 offices have? ***Ans:*** 72 chairs.

51. If one bush has 27 berries, how many berries would be on 4 bushes? **Ans:** 108 berries.

52. If a cat has 12 whiskers on each side, how many whiskers do 7 cats have on both sides? **Ans:** 168 whiskers. If 7 cats have 84 whiskers on one side then on both sides there are 168.

COUNTING MONEY

We count money by adding and subtracting dollars and cents. There are 100¢ in each dollar. We convert 2 dollars and 25 cents into 225 cents for easy calculations.

For example, if you add 50¢ and 60¢ and get 110¢, you can say that now there are 1 dollar and 10 cents. If you have 1 dollar and 10¢ and pay 95¢ for an ice cream cone, then you no longer have 110 cents, but only 15¢.

EXERCISE I

Problem: $8 and 15¢ + $7 and 35¢ = ?
Solution: First, we add dollars, 8 + 7 = 15 (dollars).
Next, we add cents, 15 + 35 = 50 (cents). Then, we put dollars and cents together, $15 and 50¢. The answer is $15 and 50¢.

Problem: $14 and 75¢ + $13 and 25¢ =?
Solution: First, we add the dollars, 14 + 13 = 27 (dollars). Next, we add the cents, 75 + 35 = 100 (cents) or $1. Then, we add sum of cents to the sum of dollars, $27 + $1 = $28.

Problem: $4 and 90¢ + $6 and 23¢ =?
Solution: First, $4 + $6 = $10; then, 90¢ + 23¢ = $1 and 13¢.

Together, $10 + $1 + 13¢ is equal to $11 and 13¢.

$1 and 50¢ + 50¢ = $2, because 50¢ + 50¢ = 100¢ or $1
$1 and 50¢ + $1 = $2 and 50¢.
$2 and 50¢ + $2 and 25¢ = $4 and 75¢

1. What is $7 and 70¢ + $3 and 25¢? *Ans:* $10 and 95¢

2. What is $10 and 25¢ + $5 and 75¢? *Ans:* $16

3. What is $2 and 90¢ + 10¢? *Ans:* $3.

4. What is $2 and 90¢ + 20¢? *Ans:* $3 and 10¢, because 90¢ + 20¢ = 110¢ or $1 and 10¢

5. What is $13 and 0¢ + $17 and 44¢? *Ans:* $30 and 44 ¢

6. What is $11 and 15 ¢ + $16 and 55¢? *Ans:* $27 and 70¢

7. What is $40 and 75¢ + $60 and 75¢? *Ans:* $101 and 50 ¢

8. What is $12 and 60¢ + 12 and 60¢? *Ans:* $25 and 20¢.

9. What is $17 and 28¢ + $13 and 39¢? *Ans:* $30 and 67¢

10. What is $35 and 70¢ + $19 and 35¢? *Ans:* $55 and 5¢

11. What is $50 and 55¢ + $10 and 55¢? *Ans:* $61 and 10¢

12. What is $95 and 90 ¢ + $10 and 90¢? *Ans:* $106 and 80¢

13. What is $18 and 25¢ + $25 and 39¢? *Ans:* $43 and 64¢

We can also subtract money by taking out first dollars, then cents.

Problem: $7 and 45¢ - $3 and 30¢ = ?
Solution: First, we subtract dollars, $7 and 45¢ - $3 = $4 and 45¢.
Then, we take away the cents, $4 and 45¢ - 30¢ = $4 and 15¢.

$7 and 45¢ - $3 and 30¢ = $4 and 15¢

Problem: $10 - $2 and 50¢ = ?
Solution: First, we subtract dollars, $10 - $2 = $8.
Then, we take away the cents, $8 - 50¢ = $7 and 50¢.
Think of this part of the problem as if there were $7 and 100¢ and we took away 50¢.

$10 - $2 and 50¢ = $7 and 50¢

Problem: $10 and 20¢ - $5 and 40¢ = ?

Solution: First, we subtract dollars, $10 and 20¢ - $5 = $5 and 20¢. Then, we take away the cents, $5 and 20¢ - 40¢ = $4 and 80¢. We had to "borrow" one dollar to take 20¢ from it and were left with one less dollar and 80¢.

$10 and 20¢ - $5 and 40¢ = $4 and 80¢

1. $100 - $50 and 50¢ = $49 and 50¢

2. What is $25 and 80¢ - $16 and 40¢? *Ans:* $9 and 40¢

3. What is $70 and 40¢ minus $20 and 17¢? *Ans:* $50 and 23¢

4. What is $15 and 73¢ minus $14 and 99¢? *Ans:* 74¢

5. What is $20 and 20¢ - $15 and 40¢? *Ans:* $4 and 80¢

WORD PROBLEMS

1. Olga had $3 and 50¢ and borrowed 75¢. How much does she have now? *Ans:* $4 and 25¢.

2. If TJ gave a $10 bill for a $1 and 65¢ birthday card, how much change did he receive? *Ans:* $8 and 35¢.

3. Mr. Louis Lumiere had $20. The movie ticket cost $9 and 50¢. How much money was left for popcorn? *Ans:* $11 and 50¢.

4. His brother Auguste bought $9 and 50¢ ticket for himself and a $4 and 25¢ ticket for his daughter Nicole. How much did he pay for both tickets? *Ans:* $13 and 75¢.

5. Cheese costs $3 and 45¢ and bread costs $2 and 70¢. How much do both cost? *Ans:* $6 and 15¢.

6. The price of a toy boat is $9 and 40¢, and a toy sailor costs $5 and 50¢. How much do you pay for both? *Ans:* $14 and 90¢.

7. A hook and a sinker cost $6 and 10¢. If the hook is $5 and 20¢, what's the price of the sinker? *Ans:* 90¢.

8. A watch with a silver chain cost $25 and 50¢. If the chain is $9 and 10¢, how much is the watch? *Ans:* $16 and 40¢.

9. A carpenter paid $19 and 90¢ for a new drill and $7 and 50¢ for the drill bits. How much was the whole purchase? *Ans:* $27 and 40¢.

    Solution: An easy way to solve it is to add 10¢ to the price of the drill, that makes it $20 even, then subtract 10¢ from the price of the drill bits, $7 and 50¢ - 10¢ = $7 and 40¢. Finally, add both together, $20 + $7 and 40¢ = $27 and 40¢.

10. One can of tuna costs $1 and 65¢.
    a) How much do two cans cost? *Ans:* $3 and 30¢.
    b) How much do three cans cost? *Ans:* $4 and 95¢

11. Ingrid paid for her new dress with a $100 bill and got back $30 and 85¢ change. How much does the dress cost?
    Ans: $69 and 15¢.

12. The gas bill was $31 and 80¢; the electric bill was $28 and 45¢. How much are both bills? *Ans:* $60 and 25¢.

13. A bunch of flowers costs $7 and 70¢. How much do two bunches cost? *Ans:* $15 and 40¢.

14. It costs $15 and 50¢ to mail a large parcel and $9 and 50¢ to send a small one. What's the difference between the prices?
    Ans: $6.

15. If an airline ticket to New Orleans costs $90 and a bus ticket is $30 and 50¢ less, how much is the bus ticket? *Ans:* $59 and 50¢.

16. A vaccine costs $12 and 90¢, and a band-aid is 30¢. What's the price of both? *Ans:* $13 and 20¢.

17. One photo album costs $9 and 45¢. What's the price of two albums? *Ans:* $18 and 90¢.

18. A cup of coffee costs $1 and 50¢ and a muffin costs 50¢.
    a) How much will you pay for one coffee and two muffins?
    Ans: $2 and 50¢.
    b) How much do two cups of coffee and one muffin cost?
    Ans: $3 and 50¢.
    c) How much are three cups of coffee? *Ans:* $4 and 50¢.
    d) How much are four muffins? *Ans:* $2.

19. A pair of gloves costs $14 and 50¢. How much does one glove cost? *Ans:* $7 and 25¢ is probably the right answer, but you can't buy only one glove.

20. A tube of toothpaste costs $2 and 10¢ and a toothbrush is 60¢ less. How much do both cost? *Ans:* $3 and 60¢.

21. One teabag at a coffee shop costs 99¢, and hot water is free. How much do three teabags cost? *Ans:* $2 and 97¢ (3¢ less than $3).

22. A DVD player costs $60 and 60¢. There is a sales tax of $5 and 20¢. What's the price with the tax? *Ans:* $65 and 80¢.

23. A mug costs $4 and 50¢. How much do three mugs cost? *Ans:* $13 and 50¢.

24. Ms. Sanchez had $64. She bought 2 movie tickets, $11 and 50¢ each. How much money does she have left? *Ans:* $41. *Solution:* The price of two movie tickets is $23. Then, $64 - $23 = $41.

25. If Trevor had $7 and 30¢ and donated $5 and 25¢ to the animal shelter, how much is left? *Ans:* $2 and 5¢.

26. Out of $99 saved for the trip, Nasser used $33 for the tickets. How much was left? *Ans:* $66.
a) Out of $66, Nasser spent $22 for the a hot air balloon ride. How much money was left? *Ans:* $44.
b) Out of $44, he spent $11 on bread to feed the ducks in the park. How much money did he have then? *Ans:* $33.
c) He paid for the $11 pony ride three times, he liked it so much. How much money was left for ice cream? *Ans:* Nothing, he spent it all.

27. Leona had $45 and 50¢ before Marc returned $15 and 55¢ that he borrowed the week before. How much money does she have now? *Ans:* $61 and 5¢.

28. A calculator costs $12 and 50¢. How much do three calculators cost? *Ans:* $37 and 50¢. Wouldn't it be easier to buy 1 calculator first and calculate the price of the other two?

29. One euro is equal to $1 and 50¢.
    a) How much are two euros? *Ans:* $3.
    b) How much are three euros? *Ans:* $4 and 50¢.
    c) How much are four euros? *Ans:* $6.

30. One British Pound is equal $1 and 70¢.
    a) How much are two Pounds? *Ans:* $3 and 40¢.
    b) How much are four Pounds? *Ans:* $6 and 80¢.

31. There are 100 kopeks in one Russian ruble. How many kopeks
    are in 3 rubles? *Ans:* 300 kopeks.
    a) If Peter the Great, the Russian Tsar, had 3 rubles and paid 1
    rubble and 13 kopeks for a new beard comb, how much did he
    get in change? *Ans:* 1 rubble and 87 kopeks.
    a) How much would he pay for 2 new combs?
    Ans: 2 rubbles and 26 kopeks.
    b) How much would he get back then? *Ans:* 74 kopeks.
    Actually, Tsar Peter was cutting beards, not combing them.

32. There are 100 new pence in one British pound sterling.
    Geoffrey had 4 pounds and paid 1 pound and 66 pence for a
    cup of tea with crumpets. How much did he get in change?
    Ans: 2 pounds and 34 pence.
    Before 1971, Britain used a different money counting system÷
    the pound was divided into 20 shillings and each shilling into
    12 pence, making 240 pence to the pound.

33. There are 100¢ in one euro. Greta paid 5 euros for a Frankfurter
    sausage and received 2 euros and 43¢ in change. What was the
    price of the sausage? *Ans:* 2 euros and 57¢.

34. Kenzie put $97 and 60¢ in the bank and earned $4 and 25¢
    interest in one year. How much money is in the bank now?
    Ans: $101 and 85¢.

35. While working in a restaurant, Lulu received $58 and 80¢ in
    salary and $40 and 50¢ in tips. How much did she earn in all?
    Ans: $99 and 30¢.

36. Mr. Jensen's bill for the dinner was $36 and 60¢. He left $5 and
    50¢ for the tip. How much did he pay? *Ans:* $42 and 10¢.

37. A newspaper subscription costs $18 and 90¢, and a magazine subscription costs $23 and 80¢. What's the price of both? *Ans:* $42 and 70¢.

38. Shoes and socks together cost $42. How much do you pay for the socks, if the shoes are $38 and 50¢? *Ans:* $3 and 50¢.

39. Pizza with toppings is $18 and 99¢. Pizza without toppings is $15 and 99¢. How much are the toppings? *Ans:* $3.

40. A dinghy with paddles costs $90. The paddles alone cost $21 and 50¢. What's the price of the dinghy without the paddles? *Ans:* $78 and 50¢.

41. Gigi's monthly allowance is $50. By the end of the month she saved $13 and 50¢. How much did she spend? *Ans:* $36 and 50¢.

42. A street juggler made $77 in tips but had to pay $16 and 50¢ fine for performing without a permit. How much money did he take home? *Ans:* $61 and 50¢.

43. This time, Red Riding Hood took a train and a bus to visit her grandmother and paid $40 and 60¢ for the trip. How much was the train ticket if she paid $10 and 50¢ for the bus? *Ans:* $30 and 10¢. She met Mr. Wolff on the train.

44. A can of white paint costs $22 and 80¢, and the red paint costs $ 4 and 20¢ less. How much does a can red pain cost? *Ans:* $18 and 60¢.
    a) How much do both cans cost together? *Ans:* $41 and 40¢.

45. A book, priced at $21 and 75¢, was sold at a discount of $2 and 50¢. What's the new price? *Ans:* $19 and 25¢.

46. Mario paid for a guitar and the strings $60 and 90¢. If the strings cost $9 and 45¢, how much was the guitar? *Ans:* $51 and 45¢.

47. Romero got paid $44 for two days of gardening. The first day he was paid $17 and 50¢. How much was the second day pay? *Ans:* $26 and 50¢.

48. Mrs. Fairchild gave $54 to her two nephews. How much money was left for Nilly if Willy got $25 and 80¢? *Ans:* **$28 and 20¢.** I don't know why she thought that was fair.

49. Timor paid for a globe with a $100 bill and got back $67 and 13¢ in change. What's the globe's price? *Ans:* **$32 and 87¢.**

50. If Dante paid $21 for two tickets, how much was each ticket? *Ans:* **$10 and 50¢.** He should have bought only one, Beatrice didn't show up.

10

ADDING, SUBTRACTING, AND CONVERTING TIME

Go over the following facts with the student and make sure he knows all these well, without a lot of hesitation.

- 60 minutes equal 1 hour.
- 120 minutes equal 2 hours.
- 180 minutes equal 3 hours.
- 240 minutes equal 4 hours.
- 300 minutes equal 5 hours.
- 360 minutes equal 6 hours.
- 600 minutes equal 10 hours
- 24 hours equal one day.
- 48 hours equal 2 days.
- 72 hours equal 3 days.
- 96 hours equal 4 days.
- 120 hours equal 5 days.
- 240 hours equal 10 days.

When you add or subtract numbers, new count starts after each ten. For example, we count 17, 18, 19, 20, then at 20 we start new set of ones, 21, 22, 23, ... We do the same with each hundred: 98, 99, 100, then 101, 102, 103...

When we add or subtract time we start new count after 60 seconds for a new minute, after 60 minutes for a new hour, and after 24 hours for every day.

For example: 58 seconds, 59 seconds, 60 seconds (or 1 minute), then 1 minute and 1 second, 1 minute and 2 seconds, 1 minute and 3 seconds, etc.

- 58 minutes, 59 minutes, 60 minutes (or 1 hour), then 1 hour 1 minute, 1 hour 2 minutes, etc.

- 22 hours, 23 hours, 24 hours (or 1 day), then 1 day and 1 hour, 1 day and 2 hours, etc.

EXERCISE I

1. How many days and hours are in 24 hours? *Ans:* 1 day only

2. How many days and hours are in 30 hours?
   Ans: 1 day and 6 hours (24 + 6)

3. How many days and hours are in 34 hours?
   Ans: 1 day and 10 hours

4. How many days and hours are in 48 hours?
   Ans: 2 days (24 + 24)

5. How many days and hours are in 50 hours?
   Ans: 2 days and 2 hours (24 + 24 + 2)

6. How many days and hours are in 60 hours?
   Ans: 2 days and 12 hours

7. How many days and hours are in 240 hours? *Ans:* 10 days

8. How many hours are in 3 days? *Ans:* 72 hours

9. How many hours are in 4 days? *Ans:* 96 hours

10. How many hours are in 5 days? *Ans:* 120 hours

11. How many hours are in one week? *Ans:* 168 hours
    (5 days + 2 days or 120 + 48)

12. How many days and hours are in 100 hours?
    Ans: 4 days and 4 hours (24 + 24 + 24 + 24 +4).

EXERCISE II

- 1 hour equals 60 minutes, then 1 hr + 60 minutes = 2 hrs
- 1 hr + 75 minutes make 2 hrs 15 min, because 75 min are equal to 60 min + 15 min
- 1 hr + 90 minutes make 2 hrs and 30 min, because 90 min equal to 60 min + 30 min

1. What is 30 min + 30 min? *Ans:* 60 minutes, or 1 hr.

2. What is 31 min + 30 min? *Ans:* 61 minutes, or 1 hr and 1 min.

3. What is 45 min + 30 min? *Ans:* 75 min, or 1 hr and 15 min.

4. What is 45 min + 45 min? *Ans:* 90 minutes, or 1 hr and 30 min (90 is equal to 60 + 30).

5. What is 60 min + 60 min? *Ans:* 120 minutes, or 2 hrs.

6. What is 90 min + 90 min? *Ans:* 180 minutes, or 3 hrs (60 + 60 + 60 = 180).

7. What is 40 min + 30 min? *Ans:* 70 minutes, or 1 hr and 10 min.

8. What is 55 min + 35 min? *Ans:* 90 minutes, or 1 hr and 30 min.

9. What is 75 min + 28 min? *Ans:* 103 min, or 1 hr and 43 min.

10. What is 59 min + 66 min? *Ans:* 125 minutes, or 2 hrs and 5 min.

EXERCISE III

1. It's 3:30 PM. What time was it 20 min ago? *Ans:* 3:10 PM

2. It's 3:15 PM. What time was it 15 min ago? *Ans:* 3:00 PM

3. It's 4:20 AM. What time will it be in 30 min? *Ans:* 4:50 AM

4. It's 1:50 PM. What time will it be in 20 min? *Ans:* 2:10 PM (10 min to 2 PM to make full 60 minutes, and 10 min after the hour)

5. It's 5:55 AM. What time will it be in 20 min? *Ans:* 6:15 AM (20 is equal to 5 + 15)

6. It's 7:05 PM. What time was it 10 min ago? *Ans:* 6:55 PM

7. It's 3:10 PM. What time was it 15 min ago? *Ans:* 2:55 PM

8. It's 9: 40 PM. What time will it be in 1 hr? *Ans:* 10:40 PM

9. It's 9:40 AM. What time will it be in 1 hr 15 min?
    Ans: 10:55 AM

10. It's 8:25 PM. What time was it 35 min ago? *Ans:* 7:50 PM (35 is equal to 25 + 10)

11. It's 9:40 PM. What time will it be in 1 hr 20 min? *Ans:* 11 PM

WORD PROBLEMS

1. It takes 23 minutes to get to the library and the same amount of time to come back. How long is the round-trip?
    Ans: 46 minutes.

2. If it took 35 minutes each way, how long would the round-trip be? *Ans:* 70 minutes or 1 hr and 10 min.

3. If mom said she'll be home in 120 minutes, how many hours is that? *Ans:* 2 hours.

4. If dad said he'll be leaving in 85 minutes, how many hours and minutes is that? *Ans:* 1 hr and 25 min.

5. It takes 180 min to make a birdhouse and 2 hrs 55 min to make a birdfeeder. Which one takes longer to make?
    Ans: the birdhouse (3 hours) but only by 5 minutes.

6. I can run 10 miles in 120 minutes, said the Roadrunner. I'll make it in 2 hours, said the Coyote. Who is faster?
    Ans: they both make in the same time.

7. Ron can fix an engine in 3 hours and Steve can do in 160 minutes. Who will finish first? *Ans:* Steve, because 160 minutes is only 2 hours and 40 minutes.

8. Jordan takes 20 minutes to shower, 20 minutes to eat breakfast, and 20 minutes to dress up for school. How much time in all? *Ans:* 60 minutes or 1 hour.

9. Teresa takes 22 minutes to shower, 22 to eat breakfast, and 22 minutes to dress up. How much time does she take?
Ans: 66 minutes or 1 hour and 6 minutes.

10. Nora left home at 7:40 AM. She walked 40 minutes to school. What time did she get there? *Ans:* 8:20 AM (40 minutes breaks into 20 minutes to make full hour, 8 AM, and another 20 minutes after).

11. How many times a day does a broken clock show the right time? *Ans:* twice.

12. Julian's job interview was scheduled at 8:55 AM. Julian called to say that he was running 15 minutes late. What time will he get there?
Ans: 9:10 AM (15 minutes splits into 5 min to make a full hour and 10 minutes after). It may not be a good idea to be late for a job interview.

13. If it takes 45 minutes for one washing cycle and 45 minutes for drying, how much time does take for the laundry? *Ans:* 90 minutes, or 1 hr and 30 min, or one and a half hour.

14. It takes 4 hours and 15 minutes to prepare and eat the turkey. It takes only 20 minutes to eat it. How long does it take to roast it? *Ans:* 3 hours and 55 minutes.

15. It takes a princess 7 hours and 5 minutes to get dressed and arrive at a costume ball. If the princess lives only 25 minutes away, how long does it take to get ready? *Ans:* 6 hours and 40 minutes. She is a princess, after all.

16. Getting ready for a surgery took 50 minutes, the surgery lasted 80 minutes, and the recovery was 70 minutes. How long did it take from the beginning to the end? *Ans:* 200 minutes or 3 hrs and 20 min (50 + 80 + 70= 200 or 60 + 60 + 60 + 20).

17. If it took 47 minutes for a pony to get up the hill and 25 minutes to get down, how long was the trip? *Ans:* 72 minutes or 1 hr and 12 min.

18. The class starts at 8:10 AM but Niraz came 20 minutes earlier. What time did he come? *Ans:* 7:50 AM.

19. Jill watches shows at 6:55 PM. Her watch is 15 minutes behind, what is the right time? *Ans:* 7:10 PM.

20. Bill watches shows at 8:20 AM but his watch is 25 minutes fast. What's the correct time? *Ans:* 7:55 AM.

21. A train was scheduled to leave at 9:45 AM but was delayed by 35 minutes, what time did the train leave? *Ans:* 10:20 AM (35 splits into 15 + 20).

22. It took the jury 24 hours to make a decision. How many days is that? *Ans:* 1 day.

23. The boy's fever lasted 48 hours. How many days? *Ans:* 2 days.

24. It was raining for 57 hours, non-stop. How many days and hours did it rain? *Ans:* 2 days and 9 hours (24 + 24 + 9).

25. The Olympic Games will start in 72 hours. How many days is that? *Ans:* 3 days.

26. It will take me 66 hours to repair the submarine's engine, said chief-mechanic. We have enough air only for 3 days, said the captain. Will the mechanic finish the repair on time? *Ans:* yes. *Solution:* The mechanic needs 66 hours, which is equal to 24 + 24 + 18, or 48 hours + 18 hours, or 2 days and 18 hours. That is less than 3 days.

27. Min Jing played 25 minutes before school and 45 minutes after. How long did she play that day? *Ans:* 70 minutes or 1 hr and 10 min.

28. A brick floated for 19 hours in the lake, then 18 hours in the river and 17 hours in the bay until it made to the ocean. How many days was that? *Ans:* bricks don't float, they sink. But if they did, it would have been 54 hours or 2 days and 6 hours.

29. A tailor took 17 hours to sew a pair of pants and 33 hours for a jacket. How much time did she take for the suit? *Ans:* 50 hours or 2 days and 2 hours.

30. A grandfather clock counts hours and strikes every hour up to 12; then, it starts again from one. How many times does the clock strike in one day? *Ans:* we need to do additions, 1 + 2 + 3 + 4... up to 12. That makes 78 for 12 hours. For 24 hours the clock strikes 156 times.

31. It took Nina 93 seconds to read a paragraph and 89 seconds to answer the questions. How long (minutes and seconds) did it take for both? *Ans:* 182 seconds or 3 minutes and 2 seconds.

32. A professional runner runs a mile in 5 minutes 59 seconds. An amateur runs it in 6 minutes 49 seconds. What's the difference? *Ans:* 50 seconds
(1 second up to 6 minutes for a professional plus 49 seconds). The world record for running 1 mile belongs to Hicham El Guerrouj "King of the Mile" and is 3 minutes and 43 seconds.

33. A honeybee flew out and came back to the hive in 1 hour and 10 minutes. If it took 30 minutes for nectar collection, how much time did the bee spend flying each direction? *Ans:* 20 minutes (70 min (total) - 30 min (for collection) = 40 min for the round-trip, or 20 min each way).

34. Silvia complained for 2 hours that her half-hour test was too long. How many minutes did she spend complaining?
Ans: 120 minutes.

35. A grasshopper can cross the field in 3 hours. A cockroach can get across in 5 hours and 45 minutes. What's the difference? *Ans:* 2 hours and 45 minutes.

36. The lecture was scheduled to end at 10:45 AM. But the speaker spent 20 minutes over to answer questions. When was it finally over? *Ans:* 11:05 AM.

37. The game would have ended at 3:50 PM if not for the 25 minute overtime. What time did it end? *Ans:* 4:15 PM.

38. To catch 8:30 AM flight, Mr. Punctual came 1 hour and 45 minutes earlier than he was supposed to. What time did he get to the airport? *Ans:* 6:45 AM.

39. A 5:40 PM flight to Puerto Rico was delayed by 1 hour and 35 minutes. What time did it take off? *Ans:* 7:15 PM

40. After the surgery, you can't take a shower for 72 hours, said the nurse. How many days is that? *Ans:* 3 days.

41. Your birthday is in 240 days. How many months is that from now? *Ans:* 6 months.

42. School ends 37 days. How many months and weeks is that from now? *Ans:* 1 month and 1 week.

43. School ends 141 days. How many months and weeks is that from now? *Ans:* 4 months and 1 weeks.

MULTIPLICATION REVIEW, SQUARES AND CUBES

When we multiply a number by itself we say the number is squared and call the product square of the number. For example: 2 × 2 = 4, or 2 square equals 4.

When we multiply the number by itself three times we call the product cube of the number. For example: 2 × 2 × 2 = 8, or 2 cube equals 8.

EXERCISE I

1. What is 3 squared? *Ans:* 9, because 3 × 3 = 9

2. What is 1 squared? *Ans:* 1, because 1 × 1 = 1

3. What is 2 squared? *Ans:* 4, because 2 × 2 = 4

4. What is 5 squared? *Ans:* 25, because 5 × 5 = 25

5. What is square of 7? *Ans:* 49, because 7 × 7 = 49

6. What is 4 squared? *Ans:* 16

7. What is 6 squared? *Ans:* 36

8. What is 10 squared? *Ans:* 100

9. What is 0 squared? *Ans:* 0, because 0 × 0 = 0

10. What is 9 squared? *Ans:* 81

11. What is 8 squared? *Ans:* 64

EXERCISE II

1. What is 3 cubed ? *Ans:* 27, because 3 × 3 = 9, and 9 × 3 = 27

2. What is 4 cube of? *Ans:* 64, because 4 × 4 = 16, then
    16 × 4 = 64

3. What is 1 cubed? *Ans:* 1, because 1 × 1 = 1, and then 1 × 1 = 1

4. What is 2 cubed? *Ans:* 8, because 2 × 2 = 4, and then 4 × 2 = 8

5. What is 5 cubed? *Ans:* 125, because 5 × 5 = 25, and then
    25 × 5 = 125 (It's hard to multiply 25 by 5, then adding 25 five
    times might work.)

EXERCISE III

2 × 5 = 10	5 × 7 = 35	7 × 8 = 56
2 × 9 = 18	5 × 5 = 25	7 × 4 = 28
3 × 7 = 21	5 × 9 = 45	8 × 4 = 32
3 × 9 = 27	6 × 3 = 18	8 × 5 = 40
4 × 3 = 12	6 × 7 = 42	8 × 9 = 72
4 × 8 = 32	6 × 9 = 54	9 × 4 = 36
4 × 7 = 28	6 × 5 = 30	9 × 6 = 54
4 × 9 = 36	7 × 6 = 42	9 × 8 = 72

▶ *A trick: Multiplying by 5.* An easier way to multiply by 5 is to divide
the number by 2 and then multiply it by 10.

Problem: 12 × 5 = ?

Solution: First, divide 12 by 2, 12 ÷ 2 = 6. Then 6 × 10 = 60

12 × 5 = 60

14 × 5 = 70

16 × 5 = 80

18 × 5 = 90

20 × 5 = 100

WORD PROBLEMS

1. Doris has 6 goldfish, and Boris has 3 times as many. How many gold fish do both have? *Ans:* 24 goldfish.

2. If parking is $3 per hour, how much will we pay for 6 hour parking? *Ans:* $18.

3. Each car in a railroad set costs $7, how much do 6 cars cost? *Ans:* $42.

4. On a trip to Alaska, Ron took 8 pictures and Dean took 7 times as many. How many pictures did Dean take? *Ans:* 56 pictures.

5. If there are 8 toads in a swamp, how many toads are in 4 swamps? *Ans:* 32 toads.

6. If one slice of pizza is $3, how much are 9 slices? *Ans:* $27.

7. If there are 5 campers in one tent, how many campers are in 7 tents? *Ans:* 35 campers.

8. The shop has 5 packs of green tea and 5 times as many of black tea. How many packs of both kinds do they have?
Ans: 30 packs.

9. A ticket to the art museum is $4 and to a baseball game is 6 times more. What's the price of the game tickets? *Ans:* $24.

10. A ticket to the art museum is $4. What's the price of 7 tickets? *Ans:* $28.

11. There are 8 campers per each counselor in the camp. How many people are at the camp if there are 6 counselors?
Ans: 54 people.
Solution: There are 6 counselors and 8 camper for each counselor. There are 8 × 6 = 48 (campers). Together, 48 (campers) + 6 (counselors) = 54 people.

12. One mailbox has 9 letters, the other 4 times as many. How many letters are in both boxes? *Ans:* 45 letters.

13. A hockey game lasts 3 periods, 20 minutes each. How long is that? *Ans:* 60 minutes or 1 hour.

14. One farm has 7 acres, the other is 7 times bigger. How big is the second farm? *Ans:* 49 acres.

15. A pony eats 8 pounds of carrots but an elephant can eat 8 times more. How many pounds of carrots do both eat? *Ans:* 72 pounds.
According to the Guinness Book of Records (1996) the World's Largest Carrot was grown by Bernard Lavery from South Wales, weighing in almost 16 pounds.

16. By recycling paper, a family can save 6 trees a year.
a) How many trees can 5 families save? *Ans:* 30 trees.
b) How many trees can 8 families save? *Ans:* 48 trees.
c) How many trees can 9 families save? *Ans:* 54 trees.
d) How many trees can 11 families save? *Ans:* 66 trees.

17. One camel can carry 5 bags of rice. How many bags of rice can a caravan of 8 camels carry? *Ans:* 40 bags.

18. A mouse is 7 centimeters long, and a rat is 6 times longer. How long is the rat? *Ans:* 42 centimeters.

19. In the Town of Catville, each person has 3 cats.
a) How many cats live with 7 Catvillians? *Ans:* 21 cats
b) How many cats live with 9 citizens of Catville? *Ans:* 27 cats

20. In Dogville City everyone has 4 dogs. How many dogs do 9 people have? *Ans:* 36 dogs.

21. Each acre in a forest has 4 ant colonies.
a) How many ant colonies are on 6 acres? *Ans:* 24 colonies.
b) How many ant colonies are on 7 acres? *Ans:* 28 colonies.
c) How many ant colonies are on 9 acres? *Ans:* 36 colonies.

22. A square has four equal sides. The perimeter of a square is the sum of all four sides. For example If one side of the square is 5 inches, then the perimeter would be 5 in. + 5 in. + 5 in. + 5 in. = 20 inches, or 5 in. × 4 = 20 inches.
a) What is the perimeter of a square with 6 feet on each side? *Ans:* 24 feet
b) What is the perimeter of a square with 8 inches on each side? *Ans:* 32 inches
c) What is the perimeter of a square with 10 miles on each side? *Ans:* 40 miles

23. The length of a piece of fabric is 8 times longer than the width. What's the length if the width is 3 feet long? *Ans:* 24 feet.

24. An actress has 8 dark hair wigs and 4 times as many blond wigs. How many wigs of both kinds does she have? *Ans:* 40 wigs.

25. The bus ticket to Washington, DC costs 9 times more than $8 ticket to the state capital. How much is the ticket to Washington, DC? *Ans:* $72

26. If 8 sisters have 8 earrings each, how many earrings do they have? *Ans:* 64 earrings.

27. If 7 brothers have 5 belt buckles each, how many buckles is that? *Ans:* 35 buckles.

28. It costs $7 to get across Central Park in a cab. It costs 6 times more to do it on a horse buggy. How much does it cost on the buggy? *Ans:* $42.

29. If Doris can catch 8 fish in 1 hour, how many can she catch in 4 hours? *Ans:* 32 fish.

30. 7 airplanes land at the airport each day.
a) How many land in one week? *Ans:* 49 planes.
b) How many land in two weeks? *Ans:* 98 planes.

31. There are 6 bags, each bag holds 6 cats. How many cats are there? *Ans:* 36 cats.

32. If there are 6 cages with 11 parrots in each cage, how many birds are there? *Ans:* 66 parrots.

33. A student hostel charges $7 per person, per night. How much will it cost for 5 people to stay 2 nights? *Ans:* $70.
a) It will cost 5 people $7 × 5 = $35 to stay one night. It will cost $35 + $35 = $70 for them to stay 2 nights.
b) The other way to solve it÷ It will cost $7 × 2 = $14 for one person to stay 2 days. There will be $14 × 5 = $70 for five people.

34. If one chipmunk has 7 stripes and there are 4 chipmunks in a family, how many stripes do 3 chipmunk families have?
Ans: 84 stripes.
Solution: One family of chipmunks has 7 (stripes) × 4 (chipmunks) = 28 stripes. Then 3 chipmunk families have 28 + 28 + 28 = 84 stripes.

35. There are 8 windows on each floor and there are 7 floors in the building. How many windows are in 3 buildings?
Ans: 168 windows.
Solution: In one building there are 8 (windows) × 7 (floors) = 56 windows. In 2 buildings, 56 + 56 = 112. In three buildings, there are 112 + 56 = 168 windows.

36. A small art gallery has 3 rooms, each room has 3 walls with paintings, there are 3 paintings on each wall. How many paintings are in the gallery? *Ans:* 27 paintings.

37. There are 4 penguins in each room, and there are four rooms. How many legs do all the penguins have? *Ans:* 32 legs.

38. There are 3 coins in each pocket, there are 3 pockets in each coat, there are 3 coats in the closet. How many coins are there?
Ans: 27 coins.
Solutions: 3 (coins) × 3 (pockets) = 9 (coins). Then, 9 (coins in all pockets of one coat) × 3 (coats) = 27 coins.

39. There were 2 cages with 2 parrots in each cage. Each parrot got 2 crackers. How many crackers did all get? *Ans:* 8 crackers.

40. There were 3 packages, each package had 3 stamps, and each stamp is $4. How much was the postage? *Ans:* $36

41. If there are 8 pocketknives with 4 blades each, how many blades do all the pocketknives have? *Ans:* 32 blades.

42. If one dishwasher shelf holds 9 items, how many items are on 5 shelves? *Ans:* 45 items.

43. There are 7 vacuum cleaners in the office building and we bought 8 bags for each vacuum. How many bags did we buy?
Ans: 56 bags.

44. If a dragon has 3 heads, how many heads do 8 dragons have? **Ans:** 24 heads. Dragons are not real, except for the dragonfly, an insect, and Komodo dragon, a giant lizard. Each has only one head.

45. If Chandra eats 4 sandwiches a day,
   a) how many will she eat in one week? **Ans:** 28 sandwiches.
   b)) how many will she eat in 3 weeks? **Ans:** 84 sandwiches.
   Solution: If she eats 28 sandwiches in one week, then in 2 weeks she'll eat 56 sandwiches, and in 3 weeks 56 + 28 = 84 sandwiches.

46. If there are 6 pieces in each bar of chocolate
   a) how many pieces are in 8 bars? **Ans:** 48 pieces.
   b) how many pieces are in 5 bars? **Ans:** 30 pieces.
   c) how many pieces are in 7 bars? **Ans:** 42 pieces.

DIVISION

Let's do division.

EXERCISE I

8 ÷ 2 = *Ans:* 4	32 ÷ 4 = *Ans:* 8	56 ÷ 8 = *Ans:* 7
16 ÷ 4 = *Ans:* 4	36 ÷ 6 = *Ans:* 6	35 ÷ 5 = *Ans:* 7
18 ÷ 2 = *Ans:* 9	36 ÷ 4 = *Ans:* 9	64 ÷ 8 = *Ans:* 8
21 ÷ 7 = *Ans:* 3	40 ÷ 5 = *Ans:* 8	72 ÷ 9 = *Ans:* 8
24 ÷ 3 = *Ans:* 8	42 ÷ 7 = *Ans:* 6	56 ÷ 7 = *Ans:* 8
24 ÷ 4 = *Ans:* 6	45 ÷ 5 = *Ans:* 9	42 ÷ 6 = *Ans:* 7
27 ÷ 3 = *Ans:* 9	49 ÷ 7 = *Ans:* 7	54 ÷ 9 = *Ans:* 6
28 ÷ 4 = *Ans:* 7	54 ÷ 6 = *Ans:* 9	72 ÷ 9 = *Ans:* 8

EXERCISE II

1. What number multiplied by 4 makes 12? *Ans:* 3

2. What number multiplied by 3 makes 15? *Ans:* 5

3. What number multiplied by 4 makes 28? *Ans:* 7

4. What number multiplied by 3 makes 27? *Ans:* 9

5. What number multiplied by 3 makes 24? *Ans:* 8

6.   What number multiplied by 8 makes 32? *Ans:* 4

7.   What number multiplied by 5 makes 35? *Ans:* 7

8.   What number multiplied by 6 makes 42? *Ans:* 7

9.   What number multiplied by 4 makes 36? *Ans:* 9

10. What number multiplied by 8 makes 64? *Ans:* 8

11. What number multiplied by 7 makes 28? *Ans:* 4

12. What number multiplied by 7 makes 56? *Ans:* 8

13. What number multiplied by 7 makes 49? *Ans:* 7

14. What number multiplied by 6 makes 54? *Ans:* 9

15. What number multiplied by 9 makes 72? *Ans:* 8

EXERCISE III

1.   What number divided by 2 gives 8? *Ans:* 16

2.   What number divided by 3 gives 8? *Ans:* 24

3.   What number divided by 4 gives 8? *Ans:* 32

4.   What number divided by 4 gives 7? *Ans:* 28

5.   What number divided by 3 gives 9? *Ans:* 27

6.   What number divided by 9 gives 4? *Ans:* 36

7.   What number divided by 7 gives 8? *Ans:* 56

8.   What number divided by 6 gives 7? *Ans:* 42

9.   What number divided by 8 gives 3? *Ans:* 24

10. What number divided by 9 gives 4? *Ans:* 36

11. What number divided by 5 gives 6? *Ans:* 30

12. What number divided by 6 gives 8? *Ans:* 48

13. What number divided by 5 gives 9? *Ans:* 45

14. What number divided by 8 gives 8? *Ans:* 64

15. What number divided by 7 gives 6? *Ans:* 42

WORD PROBLEMS

1. In 2 years Bonny grew 8 inches. How much did she grow each year? **Ans:** 4 inches.
    Solution: If Bonny grew 8 inches in 2 years, then each year she grew 8 (inches) ÷ 2 (years) = 4 inches.

2. June took 18 books off the shelf and divided them evenly in 3 boxes. How many books are in each box? **Ans:** 6 books.

3. How many eggs are in 16 ounces (one pound) of eggs if one egg weighs 2 ounces? **Ans:** 8 eggs.

4. In Antarctica 21 penguins broke into the groups of 3. How many groups did they make? **Ans:** 7 groups.

5. Thieves stole 28 silver spoons and split them equally. How many thieves were there, if each got 7 stolen silver spoons? **Ans:** 4 thieves.
    Solution: There were 28 spoons and they were divided such a way that each person has 7 spoons. If we count by seven, the number of counts will tell us the number of thieves. Instead, we know how to divide 28 (spoons) ÷ 7 (spoons for a thief) = 4 thieves. To prove that we are correct, let's multiply the number of spoons each thief has by the number of thieves, so 7 (spoons for each thief) × 4 (thieves) = 28 spoons. Correct!

6. How many hats $4 apiece, can you buy for $20? **Ans:** 5 hats.

7. A zookeeper divided a group of 24 pythons into 2 cages. How many pythons were in each cage? **Ans:** 12 pythons.
    a) She changed her mind and divided the group into 3 cages. How many pythons went in each cage? **Ans:** 8 pythons.
    b) If she changed her mind again and put them into 4 separate cages, how many were in each cage then? **Ans:** 6 pythons.
    c) Next, she did it again and divided 24 pythons into 6 cages. How many in each cage? **Ans:** 4 pythons.
    d) Again, she brought more cages and divided 24 snakes into 8 cages? How many in each cage now? **Ans:** 3 snakes.
    e) Lastly, the zoo director told her to put all 24 snakes in one cage. Oh, well!

8. For a project, a teacher bought 32 ball bearings and divided them evenly among 4 teams. How many did each team get? *Ans:* 8 ball bearings.
It is believed that Leonardo da Vinci came up with the idea of ball bearings but the first patent was issued to a Parisian bicycle mechanic Jules Suriray.

9. A man paid $27 for 9 pounds of grapes. What was the cost of one pound? *Ans:* $3.

10. A horse walked 20 miles in 4 hours. How many miles per hour? *Ans:* 5 miles per hour.
Solution: It takes a horse 4 hours to walk 20 miles. Then each hour it is 20 miles ÷ 4 (hours) = 5 miles per hour. We can check the answer, 5 miles × 4 (hours) = 20 miles.

11. If Robert counted 28 cats' legs at his aunt's house, how many cats does she have? *Ans:* 7 cats.

12. Ulrich laid 54 bricks in 6 equal rows. How many bricks are in each row? *Ans:* 9 bricks.

13. On Monday a mechanic changed all 36 tires on the cars in his garage. How many cars did he work on? *Ans:* 9 cars (most cars have 4 tires).

14. Mrs. Cordial sends 5 cards a year to each of her best friends. How many best friends does she have if she sent 40 cards? *Ans:* 8 best friends.

15. If a carpenter used 64 screws and 8 times less nails, how many nails did he use? *Ans:* 8 nails.

16. Park's rental shop rented 6 boats and 24 oars. How many oars went to each boat? *Ans:* 4 oars.

17. If 45 pills are evenly divided into 5 bottles, how many pills are in each bottle? *Ans:* 9 pills.
b) If these pills are evenly divided into 9 bottle, how many will be in each bottle? *Ans:* 5 pills.

18. A talented poet wrote 35 poems in one week, writing the same number every day. How many did he write each day? *Ans:* 5 poems.

19. A very talented composer wrote 42 songs in one week, writing the same number every day. How many did she write each day? *Ans:* 6 songs.

20. An extremely talented artist painted 49 portraits in one week, the same number of portraits every day. How many did he paint each day? *Ans:* 7 portraits.

21. A superbly talented sculptor made 63 sculptures in one week, making the same number every day. How many did she make each day? *Ans:* 9 sculptures.

22. A teenager played 70 video games in one week, playing the same number each day. Did he use his time wisely? *Ans:* What do you think?

23. A researcher studying dolphins counted 45 breaths in 5 minutes. How many breaths per minute does the dolphin make? *Ans:* 9 breaths.
    Solution: If dolphin makes 45 breaths in 5 minutes then each minute dolphin breaths 45 (breaths) ÷ 5 = 9 breaths. Number of breaths per minute is called respiratory rate. An average healthy adult human makes about 12 breaths per minute. Newborn's respiratory rate is about 44 breaths per minute.

24. A hiker walked 27 miles in 9 hours. How many miles did he walk each hour? *Ans:* 3 miles.

25. There are 9 light bulbs in each chandelier in a ball room. How many chandeliers are in the room if there are 54 light bulbs altogether? *Ans:* 6 chandeliers.

26. A designer used 56 buttons to make 7 matching jackets. How many buttons went on each jacket? *Ans:* 8 buttons.

27. How many pillows are in each hotel room if for 8 rooms they used 40 pillows? *Ans:* 5 pillows.

28. A 42 feet cord was cut into 6 even parts.
    a) What's the length of each part? *Ans:* 7 feet.
    b) If cut into 7 equal parts, how long would be each piece? *Ans:* 6 feet.

29. Today, Sonia read 56 pages. Yesterday, she read 7 times fewer. How many pages did she read yesterday? ***Ans:*** 8 pages.

30. There are 40 bananas in one box and 5 times less in the other box. How many bananas are in the second box? ***Ans:*** 8 bananas.

31. Third graders collected $54 and first graders collected 6 times less. How much money did the first graders collect? ***Ans:*** $9.

32. The oak lived 42 years and the willow 6 times less. How old is the willow? ***Ans:*** 7 years old.

33. During the Special Olympics, 56 players were divided into 8 equal teams. How many were on each team? ***Ans:*** 7 players per team.

34. In a bakery 56 bagels were evenly divided onto 7 baking sheets. How many bagels were on each sheet? ***Ans:*** 8 bagels.

35. Little Jim drew 54 squares and 7 circles. How many times more squares than circles did he draw? ***Ans:*** 8 times.

36. Little Iris is 8 years old, her grandfather is 72. How many times the grandfather is older than Iris? ***Ans:*** 9 times.

37. In one minute, a hedgehog runs 56 feet and a beetle runs 7 feet. How many times the hedgehog is faster than the beetle? ***Ans:*** 8 times.

38. There were 48 children in a movie and 6 adults. How many times more children than adults were there? ***Ans:*** 8 times.

39. Selling the same number of ticket each hour a theater sold 72 tickets in 8 hours. How many tickets were they selling each hour? ***Ans:*** 9 tickets.

40. Callie picked 33 red and 15 green apples, and then divided them evenly in 6 bags. How many apples are in each bag?
Ans: 8 apples
Solution: Altogether Callie picked 33 + 15 = 48 apples. Then she divided them all 48 (apples) ÷ 6 (bags) = 8 apples per bag.

41. Hiram divided evenly 64 ounces of milk into 4 tall and 4 short glasses. How much milk is in each glass? ***Ans:*** 8 oz.
Solution: Don't get confused by tall and short glasses, it only

matters that they can hold the same amount of milk. There are 4 (tall) + 4 (short) = 8 glasses. Then 64 oz ÷ 8 (glasses) = 8 oz in each glass.

42. At a party 21 coffees and 21 teas were divided evenly among 6 tables. How many people were at each table if everyone got a drink? **Ans:** 7 people

43. At a warehouse 35 green chairs and 37 blue chairs were evenly divided into 8 crates. How many chairs are in each crate? **Ans:** 9 chairs.

44. Two chefs made 17 sandwiches and three other chefs made 25 sandwiches. They divided evenly all the sandwiches among 7 families with 2 children each. How many sandwiches did each family get? **Ans:** 6 sandwiches.
Solution: The number of chefs or children in the families are only to confuse you. The important is the number of the sandwiches 17 + 25 = 42 (sandwiches) and the number of families 42 (sandwiches) ÷ 7 (families) = 6 sandwiches per family.

45. Lea had 12 physics and 52 math problems. She was solving the same number of problems every day for one week. How many problems a day did she do? **Ans:** 9 problems.

46. A farmer had 114 sheep and sold 58, then he divided the rest equally into 7 flocks. How many sheep are in each flock? **Ans:** 8 sheep.

47. Lara had $100 and spent $46 on groceries. With the rest of her money she bought 9 pairs of socks. How much does one pair cost? **Ans:** $6

48. Ken put together 42 pumpkin and 39 sunflower seeds and divided them evenly into 9 patches. How many seeds went to each patch? **Ans:** 9 seeds.

49. A gallon of milk and 6 cereal boxes cost $30. How much does a cereal box cost if a gallon of milk is $6? **Ans:** $4.
Solution: Milk and cereal together cost $30. Cereal alone costs $30 - $6 = $24. Then each box costs $24 ÷ 6 = $4.

50. A ladle and 6 serving spoons cost $73, but ladle alone costs $19. What's the price of one serving spoon? **Ans:** $9.

51. Bo paid for a pitcher and 8 glasses $90. One glass costs $7. What's the price of the pitcher? ***Ans:*** $34.
Solution: If one glass costs $7, then eight glasses cost $7 × 8 = $56. For the pitcher Bo paid $90 - $56 (price of 8 glasses) = $34.

PROBLEMS WITH MIXED CALCULATIONS

WORD PROBLEMS

1. A bagel has 1 hole and a pretzel has 3. How many holes do 12 bagels and 7 pretzels have? ***Ans:*** 33 holes (1 × 12 = 12 (holes); 3 × 7 = 21; then 12 + 21 = 33).

2. For a game, 14 girls and 8 boys divided into 2 teams. How many players are on each team? ***Ans:*** 11 players (14 + 8 = 22 (kids); 22 ÷ 2 = 11).

3. Third graders made 21 stuffed animals and the fourth graders made 33. All the animals were equally divided among 6 pre-schools. How many toys did each preschool get?
Ans: 9 toys (21 + 33 = 54; then, 54 ÷ 6 = 9 toys for each school).

4. Valery bought 6 boxes with 6 yogurts in each box and ate 5 yogurts at once. How many are left? ***Ans:*** 31 yogurts (6 × 6 = 36; 36 - 5 = 31)

5. The first team picked 17 pounds of dill and the second team picked 23 pounds. How many 4 pound bags does it make?
Ans: 10 bags.

6. A duck flew 9 miles a day for 4 days and 4 miles a day for 9 days. How many miles did it fly? **Ans:** 72 miles (9 × 4 = 36; 4 × 9 = 36; then, 36 + 36 = 72 miles).

7. Parker bought 9 boxes of paper towels, 6 rolls in each box. They used 14 rolls the two months. How many rolls are left? **Ans:** 40 rolls.

8. In the morning 9 crates, with 9 watermelons in each crate, were delivered to a store. By noon there were only 29 watermelons left. How many were sold? **Ans:** 52 watermelons.

9. A tailor bought three rolls of fabric, 50 feet each. She used 75 feet and can't figure out how many feet are left. Please help. **Ans:** 75 feet more.

10. A farm collected 67 chicken eggs and 13 less duck eggs. They sold 32 eggs. How many eggs remain? **Ans:** 89 eggs.
    Solution: 67 - 13 = 54 (ducks eggs), 54 + 67 = 121 (both eggs), 121 (total) - 32 (sold) = 89 (remained).

11. Delivered to a construction site÷ 46 pine planks and 12 less maple planks. Only 55 planks were used. How many weren't? **Ans:** 25 planks.
    Solution: Maple planks were 46 - 12 = 34. Together, there were 46 + 34 = 80 planks. Then, 80 - 55 (used planks) = 25 planks left.

12. When 35 boys and 37 girls divided into teams, 9 players in each team, how many teams did they make? **Ans:** 8 teams (35 + 37 = 72; then, 72 ÷ 9 = 8).

13. Manuel picked 43 pounds of apples from one tree and 21 pounds from another. He divided the apples into 8 bags. How many pounds in each bag? **Ans:** 8 lb.

14. All the toys for needy children were evenly divided into 7 boxes. Kindergartners collected 27 toys, first graders brought 25 and second graders gathered 11. How many toys were in each box? **Ans:** 9 toys (27 + 25 + 11 = 63; 63 ÷ 7 = 9).

15. There are 4 boxes with 5 melons in each box and 6 boxes with 4 melons each. How many melons are in all boxes? **Ans:** 44 melons.

16. There are 6 cookies with 6 chocolate chips each and also 4 cookies with 7 chips each. How many chips are in all cookies? *Ans:* 64 chips.

17. There are 7 bushes with 6 tomatoes on each bush and 6 bushes with 8 tomatoes each. How many tomatoes are there? *Ans:* 90 tomatoes.

18. If there are 8 six-string guitars and 5 seven-string guitars, how many strings are on all the guitars? *Ans:* 83 strings.
Glenn Haworth holds the record of stringing and tuning of 183 strings on 31 six-string guitars in one hour.

19. One small ferry boat carries 8 passengers and makes 4 trips a day. Another boat carries 9 passengers and makes 5 trips. How many passengers can both boats carry in a day? *Ans:* 77 passengers.

20. One stack of bricks has 7 layers, 7 bricks in each layer. Another has 6 layers, 6 bricks each. How many bricks are in both stacks? *Ans:* 85 bricks.

21. Ginger planted 5 rows of basil, 8 in each row. Basil planted 6 rows of ginger, 5 in each row. How many plants did both plant? *Ans:* 70 plants.

22. A bicyclist rides 7 miles an hour. A skateboarder goes 5 miles an hour. If both start at the same time in the opposite directions, how far they'll be from each other in 5 hours? *Ans:* 60 miles. *Solution:* In 5 hours a bicyclist will be 7 (miles per hour) × 5 (hours) = 35 miles away from the starting point. A skateboarder will be 5 (mph) × 5 (hours) = 25 miles away. They move in opposite directions, then 35 (miles) + 25 (miles) = 60 (miles).

23. One tourist walks 2 miles an hour, another walks 3. Walking in opposite direction, how far will they be from each other in 9 hours? *Ans:* 45 miles (18 + 27 = 45).

24. Heirloom tomatoes cost $3 per pound, cherry tomatoes cost $4 per pound. How much will you pay for 8 pounds of each? *Ans:* $56.

25. The T-shirts cost \$9 each and the shorts are \$7. How much do the uniforms cost for eight players? *Ans:* \$128 (\$72 + \$56 = \$128).

26. Taya filled 4 six-ounce jars and 6 eight-ounce jars with honey. How many ounces of honey are in all jars? *Ans:* 72 ounces (6 × 4 = 24; 8 × 6 = 48; 24 + 48 = 72).

27. There are 7 four-worker teams and 5 six-worker teams. How many people are there? *Ans:* 58 people.

28. There are 8 nuts in each bag and 4 bags together weigh 32 ounces. How much does each nut weigh? *Ans:* 1 ounce.
Solution: If four bags weigh 32 oz, then 1 bag weighs 32 ÷ 4 = 8 oz. If 8 oz bag has 8 nuts, then 8 oz ÷ 8 (nuts inside the bag) = 1 oz.

29. There are 5 rooms in a building and each room has 4 chairs, and each chair has 4 legs. How many legs do all chairs have?
Ans: 80 legs.
You might ask why we need to know the number of chair legs. To buy the exact number of hardwood floor protectors to stick on chair legs, perhaps.

30. Aricela took out 6 seeds from each of 4 pears and 8 seeds from each of 7 pears. How many seeds does she have? *Ans:* 80 seeds.

31. Lourdes took 60 ounces of water and divided into 2 pitchers. She filled 5 glasses from one pitcher. How many ounces of water are in each glass? *Ans:* 6 ounces.

32. A publishing company mailed 4 packages of books, 8 books in each package. If each book costs \$3, how much do all the books cost? *Ans:* \$96.

33. A leopard made 6 seven-foot jumps and 7 nine-foot jumps. How far did it go? *Ans:* 105 feet (7 × 6 = 42 ft; 9 × 7 = 63 ft; then, 42 + 63 = 105 ft).

34. A post office delivered 63 samples in 9 boxes. Taylor took the samples out of 3 boxes. How many samples were left in the boxes? *Ans:* 42 samples.
Solution: First, we need to find out how many samples were in each box. If there are 63 samples in 9 boxes, then each box has

$63 \div 9 = 7$ (samples). Then, 3 boxes have $7 \times 3 = 21$ samples. Finally, to find out how many samples were left, 63 (all samples sent) - 21 (samples out of 3 boxes) = 42 (samples left in boxes).

35. If 72 pumpkins were equally divided among 8 vegetable patches and I picked all the pumpkins from 5 patches, how many are still on the ground? **Ans:** 27 pumpkins ($72 \div 8 = 9$ (pumpkins on each patch); $9 \times 5 = 45$; then, $72 - 45 = 27$).

36. There were 80 pieces of jewelry equally divided into 10 display boxes. The thieves stole some jewelry and left behind 6 boxes intact. How many jewelry pieces were stolen? **Ans:** 32 pieces of jewelry.

37. A real estate office received 28 messages in the morning and 52 in the afternoon. They divided all the messages among 8 agents. How many messages did each agent take? **Ans:** 10 messages.

38. There are 4 pages with 8 stamps on each page and 6 pages with 6 stamps each. If the price of a stamp is 2¢, how much do all the stamps cost? **Ans:** 136¢ or $2 and 36¢.

39. In 8 hours, one team paved 80 yards of the road and the second team 72 yards. How many yards both teams together were paving each hour? **Ans:** 19 yards.
    Solution: Each hour the first team paved 80 (yards) ÷ 8 (hours) = 10 (yards). The second team paved 72 (yards) ÷ 8 (hours) = 9 (yards). Together, they did 10 + 9 = 19 (yards).

40. Each week, one chicken coop delivered 28 eggs, another gave 35. How many eggs were coming each day from both? **Ans:** 9 eggs.

41. One airplane took 37 passenger and the other 55. Each passenger carried 2 suitcases. How many suitcases are on both planes? **Ans:** 184 suitcases.

42. If 3 pigs have 4 piglets each and 3 hens have 6 chicks each, how many legs do all the animals have together? **Ans:** 102 legs.
Solution: That's a very long one. First, let's count each group of animals separately. There are 4 (piglets) × 3 (pigs) = 12 (piglets), That's 12 + 3 = 15 (pigs and piglets together). They have 4 feet each, or 60 legs for all them. Remember that number. There are also 6 (chicks) × 3 (hens) = 18 chicks, and 18 chicks plus 3 hens equals 21 chicks and hens together. They have 2 × 21 = 42 legs. Now, 60 (pigs and piglets legs) + 42 (hens and chickens legs) = 102 legs. Whew!

43. A brave bird flew 29 miles a day for 3 days and 31 miles a day for 5 days. How many days did it fly? **Ans:** 8 days.

44. Chocolate candies laid in 4 by 4 rows into 2 boxes. Caramel candies were laid in 3 by 3 rows in 3 boxes. How many candies were there?
Ans: 59 candies (chocolate÷ 4 × 4 = 16 in each box, and 32 in 2 boxes; caramel÷ 3 × 3 = 9 in each box, and 27 in 3 boxes. Then, 32 + 27 = 59).

45. There are 5 thrushes' nests, 6 eggs in each nest. There are 5 woodpeckers' nests, 4 eggs in each nest. How many eggs are in all nests? **Ans:** 50 eggs.

46. A jumbo egg is 4 oz and a large egg is 2 ounces. How much do 7 jumbo and 9 large eggs weigh? **Ans:** 46 oz (4 × 7 = 28 oz; 2 × 9 = 18 oz; 28 + 18 = 46 oz).

47. A vole was digging a 101 foot tunnels. If it dug 9 feet a day for 7 days, how many more feet are left to dig? **Ans:** 38 feet (9 × 7 = 63; then, 101 - 63 = 38).

48. A grocer sold 13 melons and cantaloupes for $51. If 7 melons were $3 each, what was the price of a cantaloupe? **Ans:** $5.
Solution: First, let's find out the number of cantaloupes 13 - 7 (melons) = 6. Next, the price of all cantaloupes; if the melons cost $3 × 7 = $21, then all the rest is $51 - $21 = $30. Finally, if 6 remaining cantaloupes cost $30, then each costs $5.

14

MULTIPLYING AND DIVIDING BY 10'S, 100'S

EXERCISE I

1. Count from 800 to 1000 by adding 10 (i.e. 810, 820, 830, ... 980, 990, 1000).

2. Count backward from 800 to 500 by subtracting 20 (i.e. 800, 780, 760, ... 520, 500)

3. Count backward from 600 to 210 by subtracting 30 (i.e. 600, 570, 540, 510, 480, etc.)

4. Count from 0 up to 520 by adding 40 (i.e. 0, 40, 80, 120, 160, etc.)

5. Count from 100 up to 700 by adding 50 (i.e. 100, 150, 200, 250, etc.)

6. Count from 250 up to 850 by adding 60 (i.e. 250, 310, 370, 430, etc.)

7. Count from 1000 down to 300 by subtracting 70 (i.e. 1000, 930, 860, 790, etc.)

EXERCISE II

Multiplying numbers by 10 is easy. Add one zero at the end.

Multiplying the numbers by 100 is also easy. Add two zeros at the end.

- $5 \times 10 = 50$
- $5 \times 100 = 500$
- $8 \times 10 = 80$
- $9 \times 100 = 900$
- $14 \times 10 = 140$
- $10 \times 100 = 1000$
- $22 \times 10 = 220$
- $55 \times 10 = 550$
- $101 \times 10 = 1010$
- $40 \times 00 = 4000$
- $19 \times 100 = 1900$
- $199 \times 10 = 1990$

Dividing numbers with 0 at the end by 10 is easy. Remove the zero.

Dividing numbers that end with 00 by 100 is easy. Remove two zeros at the end.

$120 \div 10 = 12$	$530 \div 10 = 53$	$1200 \div 100 = 12$
$290 \div 10 = 29$	$760 \div 10 = 76$	$5900 \div 100 = 59$
$220 \div 10 = 22$	$730 \div 10 = 73$	$9090 \div 10 = 909$
$430 \div 10 = 43$	$9900 \div 100 = 99$	$1150 \div 10 = 115$
$1,000 \div 10 = 100$	$710 \div 10 = 71$	$100 \div 100 = 1$
$990 \div 10 = 99$	$1100 \div 10 = 110$	$1000 \div 50 = 20$

EXERCISE III

Multiplying and dividing two-digit numbers with one number ending with 0.

Problem: $40 \times 7 = ?$
Solution: 40 is equal 4×10. Then, $4 \times 7 = 28$, and $28 \times 10 = 280$

$40 \times 7 = 280$

Problem: $15 \times 20 = ?$
Solution: 20 is equal 2×10; then, $15 \times 2 = 30$, and $30 \times 10 = 300$

$15 \times 20 = 300$

Problem: $150 \div 30 = ?$
Solution: 30 is equal 10×3, then $150 \div 10 = 15$ and $15 \div 3 = 5$.

$150 \div 30 = 5$

Problem: $50 \times 50 = ?$
Solution: 50 is equal 5×10, then $50 \times 5 = 250$, and
$250 \times 10 = 2500$

$50 \times 50 = 2500$

$4 \times 20 = 80$	$11 \times 20 = 220$	$200 \div 20 = 10$
$40 \times 2 = 80$	$12 \times 30 = 360$	$200 \div 40 = 5$
$5 \times 20 = 100$	$9 \times 90 = 810$	$16 \div 2 = 8$
$4 \times 80 = 320$	$3 \times 400 = 1200$	$160 \div 2 = 80$
$7 \times 80 = 560$	$60 \times 7 = 420$	$160 \div 20 = 8$
$9 \times 30 = 270$	$80 \times 6 = 480$	$100 \div 50 = 2$
$50 \times 9 = 450$	$25 \times 40 = 1000$	$360 \div 20 = 16$

▶ *A trick:* *Multiplying by 15.* When multiplying by 15, first multiply the number by 10, then add one half of the product.

Problem: $6 \times 15 = ?$
The trick: First, $6 \times 10 = 60$, then, $60 \div 2 = 30$. Finally, $60 + 30 = 90$

$6 \times 15 = 90$

Problem: 14 × 15 = ?

The trick: First, 14 × 10 = 140, then, 140 ÷ 2 = 70.
Finally, 140 + 70 = 210

14 × 15 = 210

8 × 15 = 120

10 × 15 = 150

5 × 15 = 75

12 × 15 = 180

100 × 15 = 1500

22 × 15 = 330

WORD PROBLEMS

1. Ann has $4. How many ¢ is that? *Ans:* 400¢, because there are 100 cents in each dollar.

2. If each page has 2 sides, how many sides do 100 pages have? *Ans:* 200 sides.

3. Cory is 4 years old. His grandma is 20 times older. How old is she? *Ans:* 80 years old.

4. If a guitar has 6 strings, how many strings are on 30 guitars? *Ans:* 180 strings.

5. To clean the shipyard, 120 workers were divided into teams.
a) How many 10 men teams would it make? *Ans:* 12 teams.
b) How many 20 men teams would it make? *Ans:* 6 teams.
c) How many 30 men teams would it make? *Ans:* 4 teams.
d) How many 40 men teams would it make? *Ans:* 3 teams.
e) How many 60 men teams would it make? *Ans:* 2 teams.
f) How many 120 men teams would it make? *Ans:* 1 team.

6. Gill is 2 years old. Her grand uncle is 30 times older. How old is he? *Ans:* 60 years old.

7. There were 9 plates with 10 cherries on each plate. If we ate 30 cherries, how many are left? *Ans:* 60 cherries.
Solution: There were 10 (cherries) × 9 (plates) = 90 (cherries). Then, 90 (cherries) - 30 (cherries we ate) = 60 cherries.

8. There are 12 tents at a camp; each tent is big enough for 10 campers. How many campers can stay at the camp? **Ans:** 120 campers.

9. If there are 16 ten-gallon garbage bags, how much garbage can they hold? **Ans:** 160 gallons.

10. f there are 10 sixteen-gallon recycling cans, how much recycling stuff can they hold? **Ans:** 160 gallons.

11. Training for the marathon, 7 athletes ran holding 10 lb weight in each hand. How much weight did they all carry? **Ans:** 140 lb.
    Solution: There are two ways to solve this problem.
    a) If each athlete carries 20 lb, then, 20 lb × 7 = 140 lb.
    b) If there are 14 hands holding 10 lb each, then, 10 lb × 14 = 140 lb.

12. For a show, the circus bought 3 sea lions and 20 times as many sea otters. How many sea otters did the circus buy?
    Ans: 60 sea otters. Sea otters also differ from seals in not having blubber, a thick layer fat under the skin. Instead, they use air trapped in their fur to stay warm.

13. The photo exhibition showed 160 color photos and 20 times fewer of black and white pictures. How many black and white photos were there? **Ans:** 8 photos.

14. A farmer divided 290 chickens into coops, 10 chickens in each coop. How many coops did she use? **Ans:** 29 coops.

15. One farmer has 3 goats, the other has 30 times as many. How many goats are at the second farm? **Ans:** 90 goats. They make loud bleating noises all night long.

16. If there are 40 forks on one tray and 20 times fewer on the other, how many forks are on both? **Ans:** 42 forks.
    Solution: There are 40 forks on the first tray and also 40 (forks) ÷ 20 = 2 (forks) on the second tray. There are 40 + 2 = 42 (forks) on both trays.

17. Lynn says that she has ten $10 bills. How much money does she have? **Ans:** $100.

18. If one camel has 2 humps, how many humps do 80 camels have? *Ans:* 160 humps. Camels do not store water in their humps as is commonly believed. They use humps to collect fat.

19. If one weasel has 4 legs, how many legs do 30 weasels have? *Ans:* 120 legs.

20. If one beetle has 6 legs, how many legs do 90 beetles have? *Ans:* 540 legs.

21. If one dragonfly has 4 wings, how many wings do 70 dragonflies have? *Ans:* 280 wings.

22. One box has 60 bicycle helmets.
    a) How many helmets are in 7 boxes? *Ans:* 420 helmets.
    b) How many helmets are in 8 boxes? *Ans:* 480 helmets.
    c) How many helmets are in 60 boxes? *Ans:* 3600 helmets.
    Do you always wear helmet when you bike? You should.

23. A truck delivered 180 folding chairs. We put them in 16 rows, 10 chairs in each row. How many chairs we didn't use? *Ans:* 20 chairs.
    Solution: We put 10 (chairs) × 16 (rows) = 160 chairs. Then, 180 (delivered) - 160 (used) = 20 chairs (left).

24. The warehouse had 20 lights for each isle. If there are 9 isles in the warehouse, how many lights are there? *Ans:* 180 lights.

25. If a sewing machine makes 6 stitches per second, how many stitches does it make in 30 seconds? *Ans:* 180 stitches.

26. After $5200 were evenly divided among 10 people, how much did each person get? *Ans:* $520.

27. If 80 pounds of corn were divided among 20 pigs, how much corn did each pig get? *Ans:* 4 pounds. Not much for a pig.

28. There are 10 branches on the apple tree, each branch has 4 apples.
    a) How many apples are on the tree? *Ans:* 40 apples.
    b) If each apple has 5 seeds, how many seeds are in all of them? *Ans:* 200 seeds (5 × 40 = 200).

29. If there are 8 seeds in each pear, how many seeds are in 90 pears? *Ans:* 720 seeds.

30. If a ski gondola can lift 20 skiers, how many skiers can be on 15 gondolas? *Ans:* 300 skiers.

31. If $400 is divided among 20 people, how much will each person get? *Ans:* $20.

32. I split 400 toothpicks evenly into 10 boxes. How many toothpicks will be in each box? *Ans:* 40 toothpicks.

33. If a finch flies to the nest 20 times in one hour, how many times will it fly to the nest in 8 hours? *Ans:* 160 times.

34. A wedding ring weighs 11 grams. How much would 20 wedding rings weigh? *Ans:* 220 grams.

35. It takes a ferry 4 hours to cross the channel 10 times. How long does it take for each crossing? *Ans:* 24 minutes (4 hours is equal to 60 min × 4 = 240 minutes).

36. A duck stayed under water for 10 seconds, a swan, 7 times longer. How long did swan stay under water? *Ans:* 70 seconds or 1 minute and 10 seconds.

37. It takes 4 gallons of gas for a motorcycle to reach the state line and 20 times more for a truck. How much gas does it take for a truck? *Ans:* 80 gallons.

38. It costs Arthur $40 a month for the cell phone service, how much will he pay for 8 months? *Ans:* $320
    a) For 2 months? *Ans:* $80
    b) For 9 months? *Ans:* $360
    c) For one year? *Ans:* $480 (one year is 10 months plus 2 months).

39. If one caterpillar protection tape loops 70 centimeters around a tree, how long will be 8 loops? *Ans:* 560 cm (or 5 meters and 60 centimeters). Caterpillar protection tape is wrapped around trunk of a tree to protect it against insects.

40. If there are 11 swans and 20 times as many ducks in the lake, how many birds are there?
    Ans: 231 birds, 11 swans and 220 ducks. There were also geese in the lake, but no one bothered to count them.

41. If each garland has 25 light bulbs, how many light bulbs are on 10 garlands? *Ans:* 250 light bulbs.

42. If we put 3 eardrops in each ear, how many drops will 50 kids need? *Ans:* 300 drops.

43. If a dishwasher washed 800 plates in 10 loads, how many plates did it wash in each load? *Ans:* 80 plates.

44. At a swamp, an excavator digs out 4 loads per minute. How many loads does it dig in 60 minutes? *Ans:* 240 toads. Did I say toads? Oopsy, I meant loads, not toads.

45. There are 4 teams in the camp. In each team there are 20 students and 2 leaders. How many people are in the camp? *Ans:* 88 people.

46. If there are 18 screws and 2 screwdrivers in each set, how many items are in 5 sets? *Ans:* 100 items.

47. If there are 7 shirts and 5 pants in each closet, how many pieces of cloths are in 20 closets? *Ans:* 240 pieces.

48. If there are 15 arrows and 6 bows in each set, how many items in 10 sets? *Ans:* 210 items.

49. If Jennifer printed 15 copies of an 8 page document, how many pages did she print? *Ans:* 120 pages.

50. If Trent drew a 2 mustaches and 4 horns on each of his sister's wedding picture, how many additions did he make on 40 pictures? *Ans:* 240 additions, Trent is a perfectionists and never stops before the work is done.

15

RATIOS, RATES AND UNIT PROBLEMS

We use the term ratio to compare two numbers with the same unit. With ratios we compare apples with apples, oranges with oranges, pens with pens, and pencils with pencils.

For example: When I say Ari always has 3 times more pencils that Berry, that means no matter how many pencils Berry has, Ari has 3 times as many. We say that the ratio of Ari to Berry's pencils is 3 to 1. By knowing the number of one and the ratio, we can always figure out the other.

The ratio also works the opposite way. Using the ratio and the number of pencils Berry has we can find out Ari's.

▶ **The rule:** If Ari to Berry ratio is 3 to 1, then Berry to Ari's ratio is 1 to 3.

Problem: Ian always runs 3 times farther than Tom. That means that no matter how Tom runs, Ian runs 3 times as much. In other words, Ian to Tom ratio in distance running is 3 to 1.

a) If Tom ran 2 miles, how many miles did Ian run?
Ans: 6 miles

b) If Ian ran 15 miles, how many miles did Tom run?
Ans: 5 miles.

Problem: For every 4 sneezes I make, my sister makes 16
(I made it up, I have no sister).

a) What's mine to her sneezing ratio? ***Ans:*** 1 to 4.
b) If I sneeze 8 times, how many times will my imaginary
sister sneeze? ***Ans:*** 32 times.
c) If she sneezed 12 times, how many times did I sneeze?
Ans: 3 times, because my sister's to mine sneezing ratio
is 4 to 1.

EXERCISE I

1. Kim earns $3 for each $1 Lilly makes. What's the ratio? Ans 3
    to 1. How much did Kim earn when Lilly made $9? ***Ans:*** $27 (or
    3 times more). How much did Lilly make when Kim earned $9?
    Ans: $3 (or 3 times less).

2. I learn 6 new vocabulary words while Pam learns 12, what's
    the ratio? ***Ans:*** 1 to 2, Pam always learns twice as many words
    as I do.

Ratios can be any pair or combination of numbers. It can be 2 to 5. In
this case, for every 2 of the first there are 5 of the other.

Problem: Two stores sell pet penguins with the ratio 2 to 5.
a) If the first store sold 2 penguins, how many did the
second sell? ***Ans:*** 5 penguins.
b) If the first store sold 4 penguins, how many did the
second sell? ***Ans:*** 10 penguins

Solution: For every 2 penguins in the first store, the second sells 5.
You can think of the problem in sets. If the first store
sold 2 sets of 2 penguins each, then the second store sold
2 sets of 5 penguins each, or 5 (penguins) × 2 (sets) =
10 penguins sold.

RATES

We use the term rates to compare two different kinds of numbers. For example, we use rates to measure speed when we say that the donkey walks 6 miles per hour. That means that every hour walk, it covers 6 miles and if the donkey walks 3 hours than it makes 18 miles.

With rates we can compare miles per hour, apples per orange, and pens per pencil.

For example: Lin always buys 2 oranges with every apple. That means, there will be always 2 times as many oranges as apples regardless of how many apples he buys.

Problem: Mimi has 3 times as many pencils as pens in her back pack. If there are 15 pencils in her backpack, how many pens does she have? *Ans:* 5 pens.

EXERCISE II

1. The rate of worms used to fish caught by a fisherman is 4 to 1. If she used 24 worms, how many fish did she catch? *Ans:* 6 fish.

2. The ration of light switches to electric sockets is 3 to 2. If there are 6 switches, how many sockets does it make? *Ans:* 4 sockets.

3. If the office rate of staplers to pencils is 1 to 10, how many pencils are in the office with 3 staplers? *Ans:* 30 pencils.

4. If the rate of tea to sugar is 1 cups to 3 spoons, then how many spoons of sugar are used for 7 cups of tea? *Ans:* 21 spoons.

5. If the rate of cat's paws per cat's tails is 4 to 1, then how many tails would a cat with 20 paws have? *Ans:* 5 tails.

EXERCISE III

Solving problems with rates and ratios is easier if we reduce the rate or the ration to simple units.

> *Problem:* There are 8 books equally divided on 2 shelves. How many books would be on 3 shelves?

Solution: Step 1:   Let's find the number of books on one shelf. There are 8 books on 2 shelves, then there are 4 books on one shelf, because 8 (books) ÷ 2 (shelves) = 4 (books per shelf).

           Step 2:   If there are 4 books on one shelf, then on 3 shelves, there will be 4 (books) × 3 (shelves) = 12 books.

           a) How many books will be on 5 shelves?
           Ans: 4 (books on each shelf) × 5 = 20 books
           b) How many books will be on 10 shelves?
           Ans: 4 (books on each shelf) × 10 = 40 books

Problem: There are 24 plates in 4 plate sets.
           a) How many plates are in 5 sets?
Solution: If 24 plates make 4 sets, then each set has 24 ÷ 4 = 6 plates. Then 6 (plates in each set) × 5 (sets) = 30 plates.
           b) How many plates are in 10 sets?
           Ans: 60 plates, because one set has 6 plates and 10 sets have 60.
           c) How many plates are in 3 sets?
           Ans: 18 plates, because each set has 6 plates, then 3 sets have 18.

WORD PROBLEMS

1. Two triangles together have 6 angles. How many angles do 4 triangles have? *Ans:* 12 angles.
   Solution: If 2 triangles have 6 angles, then one triangle has÷ 6 ÷ 2 = 3 angles (but you knew that already). Then, 4 triangles have÷ 3 (angles) × 4 = 12 angles.

2. If 2 shirts have 10 buttons,
   a) how many buttons are on 3 shirts? *Ans:* 15 buttons (each shirt has 5).
   b) how many buttons are on 4 shirts? *Ans:* 20 buttons.
   c) how many buttons are on 5 shirts? *Ans:* 25 buttons.

3. If 3 bicycles have 6 wheels,
   a) how many wheels are on 6 bicycles? *Ans:* 12 wheels.
   b) how many wheels are on 7 bicycles? *Ans:* 14 wheels.

c) how many wheels are on 9 bicycles? *Ans:* 18 wheels.
d) how many wheels are on 10 bicycles? *Ans:* 20 wheels.

4. If 4 houses have 20 windows, how many windows are in 7 houses? *Ans:* 35 windows (there are 20 ÷ 4 = 5 windows in one house)

5. If 2 boxes weigh 12 pounds, how many pounds are 4 boxes? *Ans:* 24 lb.
   Solution: There are two ways to solve this problem.
   a) we know that 2 boxes are 12 lb, then one box is 6 pounds. Then, 4 boxes is 6 lb × 4 = 24 pounds.
   b) if 2 boxes weigh 12 pounds, then 4 boxes weigh twice as much. Therefore, 12 lb (2 boxes) × 2 = 24 lb.

6. If there are 24 brushes in 3 equal sets, how many brushes are in 4 sets? *Ans:* 32 brushes (8 brushes in each set).

7. If there are 32 light bulbs in 4 chandeliers, how many light bulbs are in 5? *Ans:* 40 light bulbs.

8. If 6 glasses hold 48 oz of grape juice, how much juice will 8 glasses hold? *Ans:* 64 oz.

9. If 5 posters cost $35, how much do 8 posters cost? *Ans:* $56

10. If 8 CD's cost $72, how much do 3 CD's cost? *Ans:* $27

11. If 6 guitars have 42 strings, how many strings are on 4 guitars? *Ans:* 28 strings.
    Most guitars have six strings, but seven-string guitars are popular in Russia and Brazil. There are also eight, ten, and twelve-string guitars.

12. If 6 newsletters together have 48 pages, how many pages are in 5 newsletters? *Ans:* 40 pages.

13. If 6 chipmunks have 30 stripes, how many stripes are on 7 chipmunks? *Ans:* 35 stripes.

14. If 5 cats have 45 lives, how many lives do 6 cats have? *Ans:* 54 lives. We don't know from where does the expression "a cat has 9 lives" come from. Cats were sacred animals in Ancient Egypt and number nine was a lucky number but there are no documents to prove the connection.

15. If you have to measure 14 times in order to cut 7 times, how many times do you have to measure to cut 8 times? *Ans:* 16 times to measure.

16. If 7 watches have 21 hands, how many hands are on 9 watches? *Ans:* 27 hands.

17. If 6 vans can take 42 passengers, how many passengers could 7 vans take? *Ans:* 49 passengers.

18. If 7 caddies carry 49 golf clubs, how many clubs would 4 caddies carry? *Ans:* 28 clubs.

19. If Chen bought 4 flashlights for $20, how many could he buy with $30? *Ans:* 6 flashlights.
Solution: Each flashlight costs $20 ÷ 4 = $5. If Chen has $30, then he can buy $30 ÷ $5 (price of one flashlight) = 6 flashlights.

20. If 5 window washers washed 40 windows, how many washers will wash 80 windows? *Ans:* 10 washers.
Solution: Each washer washed 40 (windows) ÷ 5 (washers) = 8 windows. Then, for 80 windows we need: 80 (windows) ÷ 8 (windows each washer washes) = 10 washers.

21. If there are 32 Chinese vases in 4 boxes, how many vases are in 7 boxes? *Ans:* 56 vases.

22. If 8 pool balls weigh 48 oz, how much do 9 pool balls weigh? *Ans:* 54 oz.

23. A truck uses 5 gallons of fuel for 50 miles of road. How far can it drive on 7 gallons? *Ans:* 70 miles.
Solution: The truck uses 50 (miles) ÷ 5 (gallons) = 10 miles per each gallon. Then using 7 gallons it can go 10 (miles with one gallon) × 7 gallons = 70 miles.

24. Another truck uses 6 gallons for 48 miles. How many miles will it go on 3 gallons? *Ans:* 24 miles.

25. If 6 ounces of perfume cost $42, how much do 7 ounces of perfume cost? *Ans:* $49.

26. If 4 ounces of silver cost $36, how many ounces of silver can you buy with $54. *Ans:* 6 ounces.

27. If 9 seashells cost $72, how much do 4 seashells cost? *Ans:* $32. Expensive seashells! In the past sea shells were used in place of money. Some Native American tribes were using long and narrow tusk shell. The shell was open on both ends and could have been worn on a string.

28. If it takes 32 months for a composer to write 4 operas, how long will it take her to write 7 operas? *Ans:* 56 months.

29. If 9 baby hamsters weigh 45 grams, how much do 5 babies weigh? *Ans:* 25 grams.

30. If whale's heart makes 16 beats in 4 minutes, how many beats does it make in 8 minutes? *Ans:* 32 beats.
    Solution: There are two ways to solve this problem.
    a) Whale's heart makes 16 beats in 4 minutes, then in 1 minute it makes $16 \div 4 = 4$ beats. In 8 minutes it makes 4 (beats) × 8 = 32 beats.
    b) If in 4 minutes the heart makes 16 beats, then in 8 minutes it will be making twice as many, or $16 + 16 = 32$ beats.

31. There are 4 rows with 3 kiwi bushes in each row. Each bush has 3 kiwis. How many kiwis are there altogether? *Ans:* 36 kiwis.
    Solution: There are 3 (bushes) × 4 (rows) = 12 bushes. Then 3 (kiwis) × 12 (bushes) = 36 kiwis.

32. Miles bought 4 boxes of dumplings, 9 dumplings in each box. He cooked them all together and then divided evenly into 6 plates. How many dumplings are on each plate? *Ans:* 6 dumplings.
    Solution: There were 9 (dumplings in each box) × 4 (boxes) = 36 dumplings. They were divided into plates÷ 36 (dumplings) ÷ 6 plates = 6 dumplings in each plate.

33. In a store 6 ladies together bought 3 packs of hairpins, 10 pins in each pack. After dividing them equally, how many pins did each lady get? *Ans:* 5 pins.

34. The store sells notebook and pencil sets. Every set comes with 2 notebooks and 3 pencils. How many pencils were sold with 18 notebooks? *Ans:* 27 pencils.

Solution: First, we need to figure out how many sets were sold. If they sold 18 notebooks and each set has 2, then 18 (notebooks) ÷ 2 (notebooks in each set) = 9 sets. If each set has 3 pencils, then 3 (pencils) × 9 sets = 27 pencils.

35. Another store also sells notebooks and pencils sets. They put 3 notebooks and 3 pencils in each set. How many pencils were sold with 30 notebooks? *Ans:* 30 pencils.

36. And yet another store sells sets of 1 notebook with 5 pencils. If they sold 50 pencils, how many notebooks were sold?
Ans: 10 notebooks.

37. A bakery had "Buy 2 Get 1 Free" muffin sale. They sold 18 muffins. How many did they give away free? *Ans:* 9 muffins, because for every 2 that were sold, 1 was given away.

38. A 5 button jacket has 2 spare buttons in the breast pocket. If the store sold 9 jackets, how many buttons went with them?
Ans: 63 buttons.
Solution: It's easy. Each jacket has 5 + 2 = 7 buttons. Then, 7 (buttons with each jacket) × 9 (jackets) = 63 buttons.

39. If a goose flies 30 miles in 5 hours, how many miles will it fly in 3 hours? *Ans:* 18 miles.

40. If it takes 6 hours for a duck to fly 30 miles, how many miles can it fly in 8 hours? *Ans:* 40 miles.

41. If it takes 10 hours for a humming bird to fly 30 miles, how many miles can it fly in 3 hours? *Ans:* 9 miles.

42. If it takes 11 hours for a fly to fly 30 mile, it will be extremely tired at a the end.

43. If they need 48 light bulbs for 6 garland lights, how many light bulbs will they need for 9 garlands? *Ans:* 72 light bulbs. Garland lights are series of light connected to each other used during holidays for weddings and also many other purposes.

44. There are 24 seats in 3 rows in the theater. How many seats are in 7 rows? *Ans:* 56 seats (8 seats in each row).

45. Altogether, there are 42 windows in 6 houses. How many windows will be in 9 houses? *Ans:* 63 windows (7 windows in each house).

46. If there are 72 cookies in 8 boxes, how many cookies are in 3 boxes? *Ans:* 27 cookies (9 cookies in each box).

47. If 54 eggplants were in 9 bags, how many were in 4 bags? *Ans:* 24 eggplants.

48. There were 65 paintings in the hotel. Some were taken out and divided among 6 guest rooms, 7 pictures in each room. The rest were taken to 'the big room'. How many paintings are in 'the big room'? *Ans:* 23 paintings.
    Solution: First, let's find out how many pictures were placed in guest rooms, 7 (paintings) × 6 (small rooms) = 42 paintings. Now, in 'the big room" there are 65 - 42 = 23 paintings.

49. Half of all Cassidy's teddy bears were divided into 6 shelves, 5 teddies on each shelf. How many teddy bears does she have? *Ans:* 60 bears.
    Solution: On the shelves, there are 5 (teddies) × 6 (shelves) = 30 teddy bears. But these are only one-half of Cassidy collection. The whole set has 30 + 30 = 60 teddy bears.

50. Gordy, a trained donkey, was helping at the farm. He was picking tomatoes and putting 4 in each box. Soon, he filled 5 boxes. He also ate some tomatoes. If altogether he picked 120 tomatoes, how many never made to the boxes? *Ans:* 100 tomatoes. Maybe donkeys should not be allowed to pick tomatoes.

51. Apollo, Greek god of hunting, ordered 100 arrows from Hephaestus, Greek god of arm making. He put 7 arrows in 7 quivers and gave the rest to his twin sister Artemis, also the goddess and a hunter. Was he fair? *Ans:* Of course he was, because he gave his sister 51 arrows and kept only 49 for himself.

16

MULTIPLICATION OF TWO-DIGIT NUMBERS

By now you have memorized single-digit times table. Then, it will be easy to multiply double-digit numbers by a single digit. First, multiply tens, next multiply ones, and then add both products together.

We will start by multiplying 11 by 11
11 is equal to 10 + 1

Next, $10 \times 1 = 10$ and $1 \times 1 = 1$. Then, $10 + 1 = 11$
$11 \times 2 = 22$, because 11 is equal to 10 + 1.
Next, $10 \times 2 = 20$ and $1 \times 2 = 2$.

Together, $20 + 2 = 22$.

$11 \times 3 = 33$, because $11 = 10 + 1$. Then, $10 \times 3 = 30$ and $1 \times 3 = 3$.

Together, $30 + 3 = 33$.

$4 \times 11 = 44$
$5 \times 11 = 55$
$10 \times 11 = 110$, right?

Problem: $13 \times 3 = ?$

Solution: 13 is equal to 10 + 3. Next, $10 \times 3 = 30$ and $3 \times 3 = 9$. Together, $30 + 9 = 39$.

Problem: $14 \times 8 = ?$

Solution: 14 is equal to 10 and 4. Next, $10 \times 8 = 80$ and $4 \times 8 = 32$. Then, $80 + 32 = 112$.

▶ *A trick:* *Let me teach you to multiply by 11.* This trick works only with some numbers.

Problem: $11 \times 11 = ?$

Solution: Take number 11 and add the digit in tens to the digit in ones, $1 + 1 = 2$. Put the sum between tens and ones, we get 121. That's the answer.
$11 \times 11 = 121$.

Problem: $12 \times 11 = ?$

Solution: First, $1 + 2 = 3$. Place 3 between 1 and 2, that's 132. That's the answer.
$12 \times 11 = 132$

Problem: $13 \times 11 = ?$

Solution: First, $1 + 3 = 4$. Place 4 between 1 and 3. That's the answer.
$13 \times 11 = 143$

- What's 14×11? **Ans:** 154
- What's 15×11? **Ans:** 165
- What's 16×11? **Ans:** 176
- What's 17×11? **Ans:** 187
- What's 18×11? **Ans:** 198

Sorry, it doesn't work as easily for 19×11. By the way, the answer is 209.

EXERCISE I

Multiplying by 12:

- $12 \times 2 = 24$
- $12 \times 3 = 36$

Remember that 3×12 is also equal 36.

$4 \times 12 = 48$, because $4 \times 10 = 40$ and $4 \times 2 = 8$. Together, $40 + 8 = 48$.
$5 \times 12 = 60$, because $5 \times 10 = 50$ and $5 \times 2 = 10$. Then, $50 + 10 = 60$.
$6 \times 12 = 72$ ($6 \times 10 = 60$ and $6 \times 2 = 12$, then $60 + 12 = 72$)

- What is 7×12? *Ans:* 84
- What is 8×12? *Ans:* 96
- What is 9×12? *Ans:* 108

EXERCISE II

Problem: $14 \times 5 = ?$
Solution: 14 is equal to $10 + 4$. Next, $10 \times 5 = 50$ and $4 \times 5 = 20$. Together, $50 + 20 = 70$

Problem: $15 \times 6 = ?$
Solution: 15 is equal $10 + 5$. Next, $10 \times 6 = 60$, $5 \times 6 = 30$. Together, $60 + 30 = 90$.

Problem: $24 \times 8 = ?$
Solution: 24 is equal $20 + 4$. Next, $20 \times 8 = 160$, and $4 \times 8 = 32$. Together, $160 + 32 = 192$

$15 \times 4 = 60$	$18 \times 3 = 54$	$17 \times 9 = 153$
$15 \times 8 = 120$	$19 \times 6 = 114$	$19 \times 5 = 95$
$13 \times 9 = 117$	$15 \times 7 = 105$	$18 \times 5 = 90$
$14 \times 8 = 112$	$17 \times 5 = 85$	$16 \times 5 = 80$
$17 \times 4 = 68$	$14 \times 9 = 126$	$14 \times 5 = 70$
$14 \times 7 = 98$	$19 \times 4 = 76$	$15 \times 5 = 75$
$17 \times 7 = 119$	$18 \times 7 = 126$	
$18 \times 6 = 108$	$15 \times 9 = 135$	
$13 \times 8 = 104$	$13 \times 7 = 91$	

▶ **A trick:** There is an easy way to multiply single and double-digit *even* numbers by 5. Since 5 equals 10 ÷ 2, we can first divide number by 2 and then multiply it by 10 to get the correct answer. Look:

- 8 × 5 = 40. First, 8 ÷ 2 = 4, then 4 × 10 = 40
- 12 × 5 = 60. First, 12 ÷ 2 = 6, then 6 × 10 = 60
- 18 × 5 = 90. First, 18 ÷ 2 = 9, then 9 × 10 = 90
- 22 × 5 = 110. First, 22 ÷ 2 = 11, then 11 × 10 = 110.

EXERCISE III: SQUARES

Square of a number is the product of the number multiplied by itself. You already know squares of single digit numbers from the times table. We use square units to measure surface area.

Then,
2 squared or 2 × 2 is equal 4
5 square is 25 and 10 squared is 100.

1. What is 1 squared? ***Ans:*** 1, because 1 × 1 = 1

2. What is 3 squared? ***Ans:*** 9

3. What is 4 squared? ***Ans:*** 16

4. What is 5 squared? ***Ans:*** 25

5. What is 6 squared? ***Ans:*** 36

6. What is 7 squared? ***Ans:*** 49

7. What is 8 squared? ***Ans:*** 64

8. What is 9 squared? ***Ans:*** 81

9. What is 10 squared? ***Ans:*** 100

10. What is 11 squared?
    Ans: 121, because 11 is equal 10 + 1. Next, 11 × 10 = 110 and 11 × 1 = 11. Then, 110 + 11 = 121.

11. What is 12 squared?
    Ans: 144, because 12 × 10 = 110 and 12 × 2 = 24. Then, 120 + 24 = 144.

WORD PROBLEMS

1. Every year Riley sends Happy New Year cards to each of her 11 cousins.
   a) How many cards did she send for the last 5 years?
   Ans: 55 cards.
   b) How many cards will she be sending in the next 7 years?
   Ans: 77 cards.
   c) How many cards will she be sending in the next 10 years?
   Ans: 110 cards.

2. If one octopus has 8 arms, how many arms do 13 of them have? *Ans:* 104 arms.

3. If one car has 4 wheels, how many wheels do 18 cars have? *Ans:* 72 wheels.

4. If one camel has 2 humps, how many humps are on 39 camels? *Ans:* 78 humps.

5. If Neeta does 16 push-ups a day, how many would it be over 6 days? *Ans:* 96 push-ups (16 is equal 10 + 6. Next, 10 × 6 = 60, 6 × 6 = 36. Then, 60 + 36 = 96).

6. If she does 7 pull-ups a day, how many will she do over two weeks (14 days)? *Ans:* 98 pull-ups.

7. When Jonah said that he can do 15 pull-ups, Brad bragged that he could do 4 times as many. How many pull-ups Brad said he could do? *Ans:* 60 pull-ups. Brad likes to show off.

8. When Leah said she will read 6 books over the summer, Brad said that could he read 12 times as many. How many books did he say he will read? *Ans:* 72 books. Remember what I told you about Brad?

9. Then Rees said she can run a mile in 6 minutes and Brad said that he can run it in 16 times more minutes. Why everyone was laughing? *Ans:* Because it would make 96 minutes, or 1 hour and 36 minutes; now everyone knows that Brad brags.

10. If a tin can holds 4 peaches, how many peaches are in 13 cans? *Ans:* 52 peaches.

Playing with a kitten a girl got 14 scratches. How many scratches would three girls get?
Ans: 42 scratches.
Cat scratches can cause blisters, pain, lymph nodes swelling, and fever. The disease is called—you guessed it—Cat Scratch Disease and needs to be treated with medicine.

11. A radio show invites 6 guests each week.
    a) How many guests came to the show over 12 weeks?
    Ans: 72 guests.
    b) How many guests came to the show over 15 weeks?
    Ans: 90 guests.
    c) How many guests came to the show over 17 weeks?
    Ans: 102 guests.

12. Omar can eat a cookie in 7 bites. In how many bites will he eat 13 cookies? *Ans:* 91 bites.

13. A restaurant bought 5 china sets, 18 items in each set. How many items did they buy?
    Ans: 90 items. Remember the trick, you can solve it by dividing 18 by 2 and then multiplying the answer (9) by 10.

14. There were 8 birch trees along the road and 12 times as many pine trees. How many pine trees were there? *Ans:* 96 pines.

15. While walking, Ping skips once after every 14 regular steps. How many steps did he make if he skipped 7 times?
    Ans: 98 steps.

16. There are 3 soup spoons in Miss Hubbard's cupboard and 12 times as many small spoons. How many big and small spoons are there? *Ans:* 39 spoons, 3 soup spoons and 36 small spoons.

17. If there are 12 petals on one daisy, how many are on 4?
    Ans: 48 petals. I asked this question to my five-year-old neighbor. He said: "I don't know, but there are 2 'petals' on my bike." I hope he was joking.

18. If there are 12 crayons in one set, how many are in 7 sets?
    Ans: 84 crayons.

19. If there are 16 pirates on each ship, how many pirates are on 5 ships? *Ans:* 80 pirates.

20. How many legs do 12 spiders have? *Ans:* 96 legs.

21. A chess board has 8 squares in each of 8 rows. How many squares are on a chess board? *Ans:* 64 squares.

22. For the family reunion we booked 18 cabins on a cruise boat. How many people are coming if there will be 4 passengers in each cabin? *Ans:* 72 people.

23. There are 12 numbers (from 1 to 12) on a clock face.
    a) How many numbers are on 4 clock faces? *Ans:* 48 numbers.
    b) How many numbers are on 6 clock faces? *Ans:* 72 numbers.
    c) How many numbers are on 5 clock faces? *Ans:* 60 numbers.
    d) How many numbers are on 9 clock faces? *Ans:* 108 numbers.

24. A princess wears 3 rings on each of her fingers and toes but not on thumbs and big toes. Altogether, how many rings does she wear? *Ans:* 48 rings on 16 fingers and toes.

25. If one ship has 7 sails, how many sails are on 16 ships? *Ans:* 112 sails.

26. If after jumping from a roof one silly teenager received 15 stitches, how many stitches would 5 silly teenagers need? *Ans:* 75 stitches.

27. To paint the entire school the painters used 8 cans of purple paint and 13 times as many cans of pink paint. How many cans of both paints did they use? *Ans:* 112 cans. Pink and purple school? Interesting...

28. Mr. Kats put 4 mousetraps in front of each of 13 mouse holes. How many traps did he put? *Ans:* 52 traps.

29. If there are 2 pieces of cheese in each mousetrap, how many pieces of cheese are there? *Ans:* 104 pieces of cheese.

30. Caroline is making 15-bead necklaces and already made 7 of them. How many beads did she use? *Ans:* 105 beads.

31. If one rental station has 15 boats, how many boats are in 7 stations? *Ans:* 105 boats.

32. If one rental station has 14 water bicycles, how many are in 5 stations? *Ans:* 70 water bicycles.

33. For the school show, Tim's idea is 9 rows, 14 chairs in each row. Tammy wants 16 rows, 7 chairs each. Whose set-up has more seats? **Ans:** Tim's, in his plan there are 126 seats. but in Tammy's only 112.

34. A magpie stole 8 plastic buttons and 17 times as many steel buttons. How many steel buttons did it steal?
Ans: 136 steel buttons. Magpie is a bird from the crow family. It is very intelligent and can learn to open pet's cage door to steal their food. They can even say words like parrots do. Magpies are known to like shiny objects they pick and take to their nests.

35. A party dress had 9 rows of buttons, 13 buttons in each row. How many buttons were there? **Ans:** 117 buttons.

36. If there are 15 teeth in each comb, how many teeth are in 7 combs? **Ans:** 105 teeth.
a) How many teeth are in 8 combs? **Ans:** 120 teeth.
b) How many teeth are in 9 combs? **Ans:** 135 teeth.

37. To memorize a new word a person needs to see it 7 times. How many times does it take for 13 new words to memorize?
Ans: 91 times.

38. There are 17 fish in a fish tank. How many fish are in 9 tanks?
Ans: 153 fish.

39. If a tomcat took 12 fish from each of 9 tanks, how many fish did it take? **Ans:** 108 fish.

40. Feeling guilty, the tomcat put back 11 fish in each fish tank. How many fish did it bring back? **Ans:** 99 fish. Ridiculous! Cats would never do that.

41. If 18 birds come to the bird feeder each day, how many might come in one week? **Ans:** 126 birds.

42. If Gunter read 16 pages in one day, how many might he read in 8 days? **Ans:** 128 pages.

43. If the width of a ribbon is 5 inches and the length is 19 times longer, what is the length? **Ans:** 95 inches.

44. If one side of a square is 9 feet, what's the surface area of the square? **Ans:** 81 feet squared.

45. If the chef puts 6 tomatoes in each cucumber salad, how many tomatoes are in the 14 salads? *Ans:* 84 tomatoes, but why does she call them cucumber salads?

46. If it takes a 7 person team to publish one book, how many people does it take to publish 17 books? *Ans:* 119 people.

47. If it takes 18 people to produce a videogame, how many people does it take to make 8 games? *Ans:* 144 people.

48. If it takes 5 people to make a home movie, how many will it take to make 15 movies? *Ans:* 75 people.

49. If one remote control has 19 buttons, how many buttons do 9 remote controls have? *Ans:* 171 buttons.

MULTIPLICATION OF A TWO-DIGIT NUMBER BY A SINGLE DIGIT

EXERCISE I

$3 \times 2 = 6$	$90 \times 5 = 450$	$40 \times 4 = 160$
$30 \times 2 = 60$	$20 \times 4 = 80$	$40 \times 7 = 280$
$2 \times 5 = 10$	$20 \times 5 = 100$	$30 \times 9 = 270$
$20 \times 5 = 100$	$20 \times 6 = 120$	$90 \times 5 = 450$
$6 \times 3 = 18$	$30 \times 3 = 90$	$70 \times 7 = 490$
$60 \times 3 = 180$	$30 \times 7 = 210$	$80 \times 6 = 480$
$9 \times 5 = 45$	$20 \times 9 = 180$	$25 \times 3 = 75$

Reminder: When multiplying any two-digit number by a single-digit number follow these rules. First, multiply tens, next multiply ones, and then add the products together.

Problem: $27 \times 6 = ?$
Solution: 27 is equal 20 + 7. Next $20 \times 6 = 120$ and $7 \times 6 = 42$. Together, $120 + 42 = 162$.

Problem: $34 \times 7 = ?$
Solution: 34 is equal 30 + 4. Next $30 \times 7 = 210$ and $4 \times 7 = 28$. Together, $210 + 28 = 238$.

17 × 7 = 119	36 × 8 = 288	29 × 9 = 261
8 × 18 = 144	46 × 7 = 322	39 × 9 = 351
5 × 25 = 125	33 × 8 = 264	49 × 9 = 441
26 × 6 = 156	39 × 7 = 273	51 × 8 = 408
34 × 8 = 272	47 × 9 = 423	
43 × 8 = 344	48 × 7 = 336	

▶ *A trick: Multiplying by 4.* It's easy to multiply by 4 if first you multiply by 2 and then again by 2.

Problem: 14 × 4 = ?
Solution: 14 × 2 = 28; then, 28 × 2 = 56.

14 × 4 = 56

WORD PROBLEMS

1. Last year a young woodpecker pecked 29 holes. This year he did 3 times as many. How many? *Ans:* 87 holes.

2. If each rock climber carries 22 pounds of weight, how much weight do 4 climbers carry? *Ans:* 88 pounds.

3. At a military parade there were 4 companies with 90 soldiers in each company. How many soldiers were there?
Ans: 360 soldiers.

4. How many minutes are in 5 hours? *Ans:* 300 minutes.
a) How many minutes are in 7 hours? *Ans:* 420 minutes.
b) How many minutes are in 3 hours? *Ans:* 180 minutes.
c) How many minutes are in 8 hours? *Ans:* 480 minutes.
d) How many minutes are in 9 hours? *Ans:* 540 minutes.

5. Each one-way trip on a shuttle bus is 27 miles long. How long are 4 one-way trips? *Ans:* 107 miles.
a) How long are 4 round-trips? *Ans:* 214 miles.
b) How long are 5 one-ways? *Ans:* 135 miles.
c) How long are 5 round-trips? *Ans:* 270 miles.

6. Big Li can lift 23 pounds, but Little Lilly can lift 6 times as many. How strong is Little Lilly? *Ans:* 138 pounds strong.

7. A skilled carpenter charges $32 per hour. How much does she charge for 4 hours of work? *Ans:* $128.

8. A skilled electrician charges $37 per hour. How much does he earn for 5 hours of work? *Ans:* $185.

9. A bus carries 28 passengers, a train 5 times as many. How many passengers can go on the train? *Ans:* 140 passengers.

10. It takes 12 hours to fly to Buenos-Aires an airplane. It takes 9 times as long on a boat. How many hours does it take on a boat? *Ans:* 108 hours or 4 days and 12 hours.

11. A train ticket to the next city is $61. The ticket across the country is 5 times more. How much is the ticket?
    Ans: $305. By the way, a trip from one coast of the United State to anther is called trans-continental or intra-continental; the trip to another continent is called inter-continental.

12. The movie title has 9 letters, the original book title has 21 times as many letters. How many letters are in the title of the book? *Ans:* 189 letters.

13. Sylvia made 5 phone calls and sent 22 times more text messages. How many text messages did she send? *Ans:* 110 text messages.

14. Mr. McGregor planted 5 watermelons and 23 times as many carrots. How many carrots did he plant? *Ans:* 115 carrots.

15. Quail Lake is 33 acres. Turkey Lake is 4 times larger. How big is Turkey Lake? *Ans:* 132 acres (33 = 30 + 3; 30 × 4 = 120, 3 × 4 = 12; 120 + 12 = 132).

16. A small hotel bought 44 new bed sheets and 4 times as many towels. How many towels did it buy? *Ans:* 176 towels.

17. The hotel bought 53 knives and 4 times as many forks. How many forks did they buy? *Ans:* 212 forks.

18. If a work shift lasts 12 hours, how long do 8 shifts last?
    Ans: 96 hours.

19. How many hours are in one week?
    Ans: 168 hours, because 24 hours × 7 (days) = 168 hours.
    a) How many hours are in 5 days? *Ans:* 120 hours
    (24 × 5 = 120, use the trick).
    b) How many hours are in 3 days? *Ans:* 72 hours.
    c) How many hours are in 4 days? *Ans:* 96 hours.
    a) How many hours are in 2 days? *Ans:* 48 hours.

20. The length of a walkway is 34 times bigger than the width, and
    the width is 6 feet. What's its length?
    Ans: 204 feet (34 is equal 30 + 4, 6 × 30 = 180, 6 × 4 = 24.
    Together 180 + 24 = 204).

21. If one rod is 28 inches, how long are 5 rods? *Ans:* 140 inches.

22. An adult has 32 teeth.
    a) How many teeth do 4 adults have? *Ans:* 128 teeth.
    b) How many teeth do 5 adults have? *Ans:* 160 teeth.
    c) How many teeth do 7 adults have? *Ans:* 224 teeth.
    d) How many teeth do 6 adults have? *Ans:* 198 teeth.

23. If there are 35 jelly beans in one pack, how many are in 5
    packs? *Ans:* 175 beans.

24. If there are 25 tea bags in one box, how many are in 5 boxes?
    Ans: 125 bags.
    a) How many bags are in 6 boxes? *Ans:* 150 bags.
    b) How many bags are in 8 boxes? *Ans:* 200 bags.
    c) How many bags are in 7 boxes? *Ans:* 175 bags.
    d) How many bags are in 4 boxes? *Ans:* 100 bags.

25. If Reed can carry 27 frogs in one bucket, how many frogs can
    he carry in 7 buckets? *Ans:* 189 frogs.

26. If he puts 36 leeches in one jar, how many leeches can fit in 5
    jars? *Ans:* 180 leeches.

27. If a stagecoach traveled 15 miles per hour, how far did it travel
    in 8 hours? *Ans:* 120 miles.

28. If a carousel makes 6 full turns per minute, how many full
    turns will it make it 17 minutes? *Ans:* 102 turns.

29. If a top makes 55 spins per minute, how many spins does it make in 7 minutes? *Ans:* 385 spins.

30. If a whirling dervish makes 37 turns each minute, how many turns does he make in 6 minutes? *Ans:* 222 turns.
Whirling dervishes got their name from a "dance" and music ceremony called the Sema. During the ceremony the dervishes (religious people) spin around, or whirl.

31. If 33 snails can be placed in one paper bag, how many can fit 6 bags? *Ans:* 198 snails.

32. If one grocery basket can hold 7 pounds of produce, how many pounds can 24 baskets hold? *Ans:* 168 pounds.

33. Mr. Notes, a composer, writes 7 sheets of music in one day. How many sheets can he write in 42 days? *Ans:* 294 sheets.

34. In one hour Mr. Notes tore 8 sheets of music. How many sheets of music he could tear in 33 hours? *Ans:* 264 sheets.

35. One sketch book has 36 pages. How many pages are in 7 books? *Ans:* 252 pages.

36. A binocular is made of 47 parts. How many parts are in 2 binoculars? *Ans:* 94 parts.

37. If there are 53 players on a team, how many players are on 3 teams? *Ans:* 159 players.
a) How many players are on 4 teams? *Ans:* 212 members.
b) How many players are on 5 teams? *Ans:* 265 members.

38. If one set has 64 colored pencils, how many pencils are in 3 sets? *Ans:* 192 pencils.

39. If one box contains 74 lollypops, how many lollypops are in 4 boxes? *Ans:* 296 lollypops.

40. If $1 can buy 83 Japanese yens, how many yens can you buy for $3? *Ans:* 249 yens.

41. $1 can buy 44 Indian rupees.
a) How many Indian rupees can you buy with $4?
Ans: 176 rupees.
b) How many Indian rupees can you buy with $3?
Ans: 132 rupees.

c) How many Indian rupees can you buy with $7?
Ans: 308 rupees.
d) How many Indian rupees can you buy with $6?
Ans: 264 rupees.

42. A safari guide told us there are 58 zebras in the park and then asked how many legs do these zebras have. Can you tell? *Ans:* 232 legs.

43. There are 33 raisins in a bag. How many raisins are in 7 bags? *Ans:* 231 raisins.

44. There are 27 nuts in a plastic bag. How many nuts are in 8 bags? 216 nuts.

45. A king invited 7 friends for hunting. Each friend brought 7 dogs. Each dog brought 7 fleas. How many fleas came? *Ans:* 343 fleas. It's easy÷ 7 (dogs) × 7 (friends) = 49 dogs, then 7 (flees) × 49 (dogs) = 343 flees. Ghastly!

46. If a trash bag holds 23 empty soda cans, how many cans would 7 bags hold? *Ans:* 161 cans. Recycle!
a) How many cans would 8 bags hold? *Ans:* 184 cans. Recycle!!
b) How many cans would 9 bags hold? *Ans:* 207 cans. Recycle!!!

47. One hair comb has 29 teeth. How many teeth do 8 combs have? *Ans:* 232 teeth. But have no fear, the combs don't bite.

48. If each sudoku puzzle has 9 squares, how many squares are in 74 puzzles? *Ans:* 666 squares.

49. If it took 53 minutes for a dog to chase a squirrel, how long would 7 dogs chase that squirrel? *Ans:* 53 minutes, dog probably chased together.

50. On a holiday, the grandpa gave $45 to each of his 7 grandchildren. How much money did he give away? *Ans:* $315.

18

MULTIPLICATION OF SINGLE DIGIT BY A TWO-DIGIT NUMBER

Remember the trick of multiplying even numbers by 5. You can divide the number by 2 and then multiply it by 10.

▶ *A trick:* *Multiplying by 15.* To multiply a single-digit number by 15, multiply the number by 10 first and then add one-half of the product.

Problem: $8 \times 15 = ?$
Solution: First, $8 \times 10 = 80$. Next, one-half of 80 is 40.
Then, $80 + 40 = 120$

$8 \times 15 = 120$

You can also add one-half of an even number to the number and then multiplying the sum by 10.

Problem: $8 \times 15 = ?$
Solution: One-half of 8 is 4 and $8 + 4 = 12$. Then $12 \times 10 = 120$.

$2 \times 15 = 30$

$5 \times 15 = 75$

$$4 \times 15 = 60$$

$$7 \times 15 = 105$$

$$8 \times 15 = 120$$

$$3 \times 15 = 45$$

$$9 \times 15 = 135$$

$$10 \times 15 = 150$$

EXERCISE I

$11 \times 5 = 55$	$14 \times 5 = 70$	$22 \times 5 = 110$
$12 \times 5 = 60$	$17 \times 5 = 85$	$30 \times 5 = 150$
$15 \times 4 = 60$	$15 \times 6 = 90$	$33 \times 5 = 165$
$16 \times 5 = 80$	$19 \times 5 = 95$	$25 \times 5 = 125$
$15 \times 5 = 75$	$15 \times 7 = 105$	$15 \times 9 = 135$
$25 \times 3 = 75$	$15 \times 8 = 120$	$25 \times 8 = 200$
$18 \times 5 = 90$	$20 \times 5 = 100$	$24 \times 5 = 120$
$13 \times 5 = 65$	$21 \times 5 = 105$	$28 \times 5 = 140$

EXERCISE II

$30 \times 5 = 150$	$64 \times 5 = 320$	$29 \times 5 = 145$
$31 \times 5 = 155$	$33 \times 5 = 165$	$35 \times 6 = 210$
$40 \times 5 = 200$	$44 \times 5 = 220$	$49 \times 5 = 245$
$42 \times 5 = 210$	$55 \times 5 = 275$	$35 \times 5 = 175$
$50 \times 5 = 250$	$66 \times 5 = 330$	$35 \times 9 = 315$
$54 \times 5 = 270$	$37 \times 5 = 185$	$46 \times 5 = 230$
$60 \times 5 = 300$	$48 \times 5 = 240$	$55 \times 5 = 275$

WORD PROBLEMS

1. Arthur does volunteer work 5 days each month. How many days does he volunteers in one year? *Ans:* 60 days ($5 \times 12 = 60$).

2. To the city dump, 17 trucks brought 5 tons of garbage each. How many tons of garbage was taken to the dump? *Ans:* 85 tons.

3. A school orchestra bought 6 new uniforms, $15 each. What was the total cost? *Ans:* $90.

4. How many months are in 5 years? *Ans:* 60 months.

5. How many months are in 7 years? *Ans:* 84 months.

6. How many months are in 4 years? *Ans:* 48 months.

7. If a dolphin takes 13 breaths a minute, how many breaths does it take in 5 minutes? *Ans:* 65 breaths.

8. If a printer prints 17 pages a minute, how many pages will it print in 5 minutes? *Ans:* 85 pages.

9. If a copier copies 15 pages a minute, how many pages will it copy in 5 minutes? *Ans:* 75 pages.

10. If a shredder cuts 19 pages a minute, how many pages does it shred in 5 minutes? *Ans:* 95 pages.

11. For her first play an actress had to memorize 23 lines. In the next play she has 5 times as many. How many lines does she have? *Ans:* 115 lines.

12. There are 25 pins on one bulletin board and 9 times more on the other. How many pins are on the second board? *Ans:* 225 pins.

13. The first day Wooley Adden's new movie played in 38 cities and 5 times in as many towns. How many towns watched the new movie? *Ans:* 190 towns.

14. Mr. Boldman counted 44 hairs on his head and Mr. Boldrick counted 5 times as many on his. How many hairs are Mr. Boldrick's head? *Ans:* 220 hairs.
Did you know that Latin word for boldness is alopecia?

15. A 27 mile road has 5 road signs for every mile. How many road signs does it have? *Ans:* 135 signs.

16. If a spider catches 7 flies each month, how many flies would 65 spiders catch? *Ans:* 455 flies.

17. In one can holds 8 oz of juice, how many oz are in 15 cans?
    Ans: 120 oz
    a) How many ounces are in 25 cans? *Ans:* 200 oz.
    b) How many ounces are in 35 cans? *Ans:* 280 oz.
    c) How many ounces are in 45 cans? *Ans:* 360 oz.
    d) How many ounces are in 55 cans? *Ans:* 440 oz.

18. If one elephant has 4 legs, how many legs do 25 elephants have? *Ans:* 100 legs. These legs are enormous.

19. A crocodile's tail is 6 feet long. How long would be the tail if the crocodile grows 35 times bigger? *Ans:* 210 feet.
    It's good that crocodiles never grow that big.

20. Each car has 4 regular wheels and one spare. How many wheels do 35 cars have? *Ans:* 175 wheels. Don't forget the spare!

21. One quarter is worth 25¢.
    a) What are 5 quarters worth? *Ans:* 125¢ or $1 and 25¢.
    b) What are 7 quarters worth? *Ans:* 175¢ or $1 and 75¢.
    c) What are 8 quarters worth? *Ans:* 200¢ or $2 even.
    d) What are 9 quarters worth? *Ans:* 225¢ or $2 and 25¢.
    e) What are 6 quarters worth? *Ans:* 150¢ or $1 and 50¢.

22. If one bike weighs 36 pounds, how much do 5 bikes weigh?
    Ans: 180 pounds.

23. If 21 Dalmatian mommies had 5 puppies each, how many puppies do all have together? *Ans:* 105 Dalmatian puppies (four puppies too many for a story).

24. One Dalmatian puppy weighs 5 pounds.
    a) How much do 27 puppies weigh? *Ans:* 135 pounds.
    b) How much do 34 puppies weigh? *Ans:* 170 pounds.
    c) How much do 42 puppies weigh? *Ans:* 210 pounds.
    d) How much do 50 puppies weigh? *Ans:* 250 pounds.

25. Whenever the brother asks Tessa for a favor she gives 5 good excuses why she couldn't do it.
    a) How many excuses would she give to 24 requests for a favor? *Ans:* 120 excuses.
    b) How many excuses would she give to 66 requests for a favor? *Ans:* 330 excuses.

26. A motorcycle has a 63 horse-power engine. How much horse-power a 5 times more powerful truck will have? *Ans:* 315 horse-powers. There are no horses inside the engine but for many years we use horse-power to measure strength of a vehicle.

27. Tamara made 5 necklaces, 74 beads in each necklace. How many beads in all? *Ans:* 370 beads.

28. Rover's doghouse is 5 feet tall. The Capitol building in Washington, DC is 58 times taller. How tall is the Capitol building? *Ans:* 290 feet.
   According to a popular myth, the D.C. laws do not allow any structures taller than the Capital building. That is not so. For example, the Old Post Office Building is 319 feet high.

29. If one Olympic symbol has 5 rings, how many rings are on 45 Olympic symbols? *Ans:* 225 rings.

30. How many pieces are in five 99 pieces puzzle? *Ans:* 495 pieces.

31. If each flight of stairs has 29 steps, how many steps are in 5 flights? *Ans:* 145 steps.

32. One piano has 88 keys. How about 5 pianos? *Ans:* 440 keys.

33. If one dinosaur laid 47 eggs, how many eggs did 5 dinosaurs lay? *Ans:* 235 eggs.

34. If a fish has 5 spines in its back fin, how many spines are in 43 fins? *Ans:* 215 spines.

35. Mr. Franco Bollo evenly divided his collection of stamps into 5 albums, 65 stamps in each album. How many stamps are in his collection? *Ans:* 325 stamps.

36. Kenny Cohen's collection of coins is divided into 48 boxes, 5 coins in each box. How many coins are in Cohen's collection? *Ans:* 240 Cohens... I mean coins.

37. If a grocery bag can fit 5 boxes of cereal, how many boxes will fit into 47 bags? *Ans:* 235 boxes.

38. If Tia puts 9 pins in her hair every morning, how many is it in 35 days? *Ans:* 315 pins (35 = 30 + 5; 9 × 30 = 210 and 9 × 5 = 45; then 210 + 45 = 315).

39. Sinbad-the-sailor promised the king to return from his voyage in 55 days but instead traveled 6 times longer. How long? **Ans:** 330 days, almost a year.

40. An alteration shop repaired 7 officer uniforms and 35 times as many soldier uniforms. How many soldier uniforms did they repair? **Ans:** 245 uniforms.

41. During a cold winter 46 field mice hid in one barn and 5 times as many in the other. How many mice hid in the second barn? **Ans:** 230 mice.

42. In one hour a bicycle can travel 35 miles and an airplane can fly 8 times as fast. How far can a plane fly in one hour? **Ans:** 280 miles (not by itself, with a pilot).

43. Uncle Vanya has 63 rubles and Three Sisters have 6 times as many. How much money do Three Sisters have? **Ans:** 378 rubles.

44. A starfish has 5 rays or arms.
    a) How many arms do 62 starfish have? **Ans:** 310 arms.
    b) How many arms do 61 starfish have? **Ans:** 305 arms.
    c) How many arms do 65 starfish have? **Ans:** 325 arms.
    d) How many arms do 64 starfish have? **Ans:** 320 arms.

45. An earthworm is 5 inches long and a fish line is 83 times longer. How long is the fish line? **Ans:** 415 inches.

46. A light breeze is 5 miles an hour; the strongest wind was 51 times faster. How fast was the strongest wind? **Ans:** 255 miles an hour.

47. An iceberg has 25 feet above the water and 6 times as much under water. What is the total height of the iceberg? **Ans:** 175 feet (under water is 25 × 6 = 150 feet; plus 25 feet above the water. Together, 150 + 25 = 175 feet).

48. An apple weighs 7 ounces, a pumpkin 65 times more. How much do both weigh? **Ans:** 462 ounces (pumpkin weighs 7 oz × 65 = 455 oz; then, 455 oz + 7 oz = 462 oz).

49. A baby dolphin is 35 inches and an adult dolphin is 5 times longer. By how many inches an adult is longer than a baby? **Ans:** by 140 inches (an adult dolphin is 35 in × 5 = 175 in; the difference is 175 in - 35 in = 140 inches).

50. A 5 milligram ant carried a twig 36 times its weight. How big was the twig? **Ans:** 180 milligrams.

51. There are 9 potatoes in a stew and 55 times as many peas. How many peas are in the stew? **Ans:** 495 peas.

52. The distance between two cities is 45 miles.
    a) How many miles is a round trip? **Ans:** 90 miles.
    b) How many miles are 5 round trips? **Ans:** 450 miles.

19

DIVISION OF TWO-DIGIT NUMBERS BY A ONE-DIGIT NUMBER

QUICK REVIEW

- *Dividend* is the number being divided.
- *Divisor* is the number that the dividend is divided by.
- The answer is called *quotient*.

In a problem $10 \div 2 = 5$, the dividend is 10 and the divisor is 2. The quotient is 5.

In $24 \div 8 = 3$, the dividend is 24, the divisor is 8, and the quotient is 3.

- What are the dividend, the divisor, and the quotient in $54 \div 9 = 6$?
- What are the dividend, the divisor, and the quotient in $54 \div 6 = 9$?
- What are the dividend, the divisor, and the quotient in $56 \div 4 = 14$?

In this lesson we will learn how to divide double-digit number larger than ten times the dividend. Division is the way to figure out how many times the divisor fits inside the dividend. For example in $36 \div 3$ we find how many 3's are inside 36.

In Verbal Math, doing it in our heads without paper and pencil, we look at the divisor as a sum of two or more numbers (try to have at least one number equal to divisor times either 10 or 20). Next, we divide each number separately and then add the quotients together. Let me show you how.

Problem: 36 ÷ 3 = ?
Solution: 36 is equal 30 + 6. Next, 30 ÷ 3 = 10 and 6 ÷ 3 = 2. Then, 10 + 2 = 12.

  36 ÷ 3 = 12

Problem: 56 ÷ 4 = ?
Solution: 56 = 40 + 16. Next, 40 ÷ 4 = 10 and 16 ÷ 4 = 4. Then, 10 + 4 = 14.

  56 ÷ 4 = 14

Problem: 65 ÷ 5 = ?
Solution: 65 = 50 + 15. Next, 50 ÷ 5 = 10 and 15 ÷ 5 = 3. Then 10 ÷ 3 = 13

  65 ÷ 5 = 13

EXERCISE I

20 ÷ 2 = 10	33 ÷ 3 = 11	64 ÷ 4 = 16
20 ÷ 10 = 2	36 ÷ 3 = 12	52 ÷ 4 = 13
22 ÷ 2 = 11	39 ÷ 3 = 13	55 ÷ 5 = 11
28 ÷ 2 = 14	40 ÷ 4 = 10	60 ÷ 5 = 12
30 ÷ 2 = 15	44 ÷ 4 = 11	65 ÷ 5 = 13
36 ÷ 3 =12	48 ÷ 4 = 12	75 ÷ 5 = 15
30 ÷ 3 = 10	55 ÷ 5 = 11	36 ÷ 3 = 12

EXERCISE II

40 ÷ 2 = 20	42 ÷ 3 = 14	60 ÷ 4 = 15
44 ÷ 2 = 22	45 ÷ 3 = 15	68 ÷ 4 = 17
48 ÷ 2 = 24	48 ÷ 3 = 16	76 ÷ 4 = 19
32 ÷ 2 = 16	54 ÷ 3 = 18	60 ÷ 5 = 12
36 ÷ 2 = 18	57 ÷ 3 = 19	75 ÷ 5 = 15
38 ÷ 2 = 19	52 ÷ 4 = 13	80 ÷ 5 = 16

90 ÷ 5 = 18	84 ÷ 7 = 12	88 ÷ 8 = 11
72 ÷ 6 = 12	91 ÷ 7 = 13	96 ÷ 8 = 12
90 ÷ 6 = 15	98 ÷ 7 = 14	78 ÷ 3 = 26

EXERCISE III

38 ÷ 2 = 19	69 ÷ 3 = 23	75 ÷ 3 = 25
40 ÷ 2 = 20	80 ÷ 4 = 20	81 ÷ 3 = 27
42 ÷ 2 = 21	84 ÷ 4 = 21	87 ÷ 3 = 29
44 ÷ 2 = 22	48 ÷ 2 = 24	88 ÷ 4 = 22
57 ÷ 3 = 19	50 ÷ 2 = 25	92 ÷ 4 = 23
60 ÷ 3 = 20	54 ÷ 2 = 27	90 ÷ 3 = 30
66 ÷ 3 = 22	45 ÷ 3 = 15	42 ÷ 3 = 14

WORD PROBLEMS

1. There are 32 teeth in an adult's mouth equally divided between upper and lower jaws. How many teeth are in each jaw? *Ans:* 16 teeth.

2. Every night old McGregor divides his 26 false teeth evenly between 2 glasses. How many teeth does he put in each glass? *Ans:* 13 teeth.

3. A school paper article said: "42 eyes looked at the substitute PE teacher". How many students looked at her? *Ans:* 21 students.

4. The coach bought 36 shoulder guards for the team. How many players are on the team? *Ans:* 18 players.

5. If all the cows on a pasture together have 56 feet, how many cows are there? *Ans:* 14 cows.

6. A 36-student school marching band marches in 3 equal columns. How many band players were in each column? *Ans:* 12 players.

7. Rodney had to write "I am sorry" 33 times. How many words does it make? *Ans:* 99 words, because "I am sorry" has 3 words.

8. If a ferry made 66 trips in one month, how many round-trips did it make? *Ans:* 33 round-trips.

9. How many triangles together have 36 angles? ***Ans:*** 12 triangles.

10. How many squares together have 44 angles? ***Ans:*** 11 squares.

11. How many octopi together have 96 legs? ***Ans:*** 12 octopi (96 is equal to 80 + 16. Next, 80 ÷ 8 = 10 and 16 ÷ 8 = 2. Finally, 10 + 2 = 12).

12. How many three-headed dragons together have 45 heads? ***Ans:*** 15 dragons, but I wouldn't stay away even from a one-headed.

13. If the length of one log is 6 feet, how many logs can you fit end-to-end into a 78 feet ditch? ***Ans:*** 13 logs (a hint: 78 is equal 60 + 18).

14. If a sari is 5 yards long, how many saris can you make out of 75 yards of fabric? ***Ans:*** 15 saris. Sari, a type of traditional women's cloth in India, is a colorful fabric wrapped in a special way around the body.

15. How many 5 flower bunches can you make from 85 flowers? ***Ans:*** 17 bunches.

16. How many two-hour classes can be scheduled in 24 hours? ***Ans:*** 12 classes.

17. When 66 daily chores were evenly divided among 3 brothers, how many chores did each get? ***Ans:*** 22 chores each. Do you think it's fair?

18. Four fishermen caught 60 fish and divided them equally. How many fish did each get? ***Ans:*** 15 fish.

19. A grandma made 48 dumplings to feed 4 grandchildren. How many dumplings were for each kid? ***Ans:*** 12 dumplings.

20. There are 65 novels in a cabinet equally divided among 5 book shelves. How many novels are on each shelf? ***Ans:*** 13 novels.

21. If a printer prints 39 pages in 3 minutes, how many pages does it print in one minute? ***Ans:*** 13 pages.

22. If a chef prepared 64 dishes during her 4 hour shift, how many dishes was she making each hour? ***Ans:*** 16 dishes.

23. If a juice-squeezer squeezes 78 oranges in 6 minutes, how many does it squeeze in one minute? *Ans:* 13 oranges.

24. If during his practice an ice skater jumped 42 times in 3 minutes, how many jumps did he do each minute?
Ans: 14 times.

25. A 60 feet tree trunk was cut into 5 equal pieces. How long was each piece? *Ans:* 12 feet.

26. A small river cruise ship has 34 cabins equally divided between 2 decks. How many cabins are on each deck? *Ans:* 17 cabins.

27. There are 39 passengers on a train, 3 passengers in each compartment. How many compartments are occupied? *Ans:* 13 compartments.

28. Moving to a new locating a dealership was loading 4 motorcycles on each truck.
a) How many truckloads will it take to move 52 motorcycles?
Ans: 13 truckloads.
b) How many truckloads will it take for 64 motorcycles?
Ans: 16 truckloads.
c) How many truckloads will it take for 68 motorcycles?
Ans: 17 truckloads.
d) Will it take more or fewer trips if they load 6 bikes on each truck? *Ans:* fewer trips.

29. If the fine for littering in the park is $45 and for walking on grass is 3 times less, how much is "walking on grass" fine?
Ans: $15.

30. The fine for "feeding the pigeons" in the park is 5 times bigger than the fine for "walking on grass". How much is it? *Ans:* $75.

31. At a gas station Mr. Ben Zinn paid $36 to fill the tank at a price $3 per gallon. How many gallons of fuel did he buy?
Ans: 12 gallons.

32. Mr. Gas O'Lin bought 42 gallons of gas and divided it evenly into 3 canisters. How much in each canister? *Ans:* 13 gallons.

33. If the rain season lasts 84 days, how many weeks is that?
Ans: 12 weeks (a hint: 84 is equal 70 + 14).

34. How many weeks is 98 days? ***Ans:*** 14 weeks (A hint: 98 is equal 70 + 28).

35. During an archery competition 72 arrows were divided into 6 quivers. How many arrows in each? ***Ans:*** 12 arrows.

36. At a car factory, 68 tires were mounted in one hour. How many cars got their tires? ***Ans:*** 17 cars, you remembered that each car has 4 tires and we didn't count the spares.

37. On a field trip, 56 second graders slept in 4 cabins. How many were in each cabin? ***Ans:*** 14 students.

38. In the desert 3 hungry explorers caught 45 lizards and divided them evenly. How many lizards did each get? ***Ans:*** 15 lizards, they didn't eat them if that what you think.

39. Charles divides 70 monkeys into 5 cages. How many monkeys went into each cage? ***Ans:*** 14 monkeys.

40. How many horses can you shoe with 56 horseshoes?
    Ans: 14 horses.

41. How many 7 apples bags can you fill with 91 apples?
    Ans: 13 bags.

42. How many $5 tickets can you buy for $85? ***Ans:*** 17 tickets.

43. How many $5 tickets can you buy for $95? ***Ans:*** 19 tickets.

44. How many $4 tickets can you buy for $48? ***Ans:*** 12 tickets.

45. How many $3 tickets can you buy for $48? ***Ans:*** 16 tickets.

46. How many $2 tickets can you buy for $48? ***Ans:*** 24 tickets.

47. Mr. and Mrs. Fox divided 76 wedding pictures into 4 albums. How many photos are in each album? ***Ans:*** 19 photos.

48. A store cut a 65-inch long sandwich into 5 pieces. How long was each piece? ***Ans:*** 13 inches. The longest sandwich recorded in Guinness book of World Records was 4,521 feet.

49. A ship with 33 crew and 63 passengers carries 8 lifeboats on board. How many people can for each lifeboat?
    Ans: 12 people (33 + 63 = 96, 96 ÷ 8 = 12).

50. If 51 fans sat on 3 benches, how many sat on each bench?
    Ans: 17 fans.

51. If 57 fans sat on 3 benches, how many sat on each bench?
    Ans: 19 fans.

52. If 76 fans sat on 4 benches, how many sat on each bench?
    Ans: 19 fans.

53. If 95 fans sat on 5 benches, how many sat on each bench?
    Ans: 19 fans.

20

DIVISION OF TWO AND THREE DIGIT NUMBERS BY ONE-DIGIT NUMBER

It's easy to divide a double-digit number by a single-digit one if we can turn it into two numbers.

▶ *The rule:* Doing division we split the number into two or more parts and divide each part separately. Then, we add our results together.

Problem: 78 ÷ 6 =?
Solution: 78 is equal to 60 + 18. Next, 60 ÷ 6 = 10 and 18 ÷ 6 = 3. Then, 10 + 3 = 13.
78 ÷ 6 = 13

Problem: 60 ÷ 3 = ?
Solution: 60 is equal to 30 + 30. Next, each 30 ÷ 3 = 10. Then, 10 + 10 = 20.

60 ÷ 3 = 20

We can also make a sum of more than two numbers. Try to see how many "ten times the divisor" numbers are within the number we divide.

Problem: 90 ÷ 3 = ?

Solution: 90 is equal to 30 + 30 + 30. Next, we divide each 30 by 3, 30 ÷ 3 = 10.

Then, 10 + 10 + 10 = 30.

90 ÷ 3 = 30

Problem: 92 ÷ 4 = ?

Solution: 92 is equal to 40 + 40 + 12. Next, each 40 ÷ 4 = 10 and 12 ÷ 4 = 3.

Then, 10 + 10 + 3 = 23

92 ÷ 4 = 23

EXERCISE I

33 ÷ 3 = 11	48 ÷ 2 = 24	60 ÷ 6 = 10
66 ÷ 3 = 22	36 ÷ 2 = 18	68 ÷ 4 = 17
69 ÷ 3 = 23	58 ÷ 2 = 29	88 ÷ 4 = 22
60 ÷ 5 = 12	40 ÷ 2 = 20	96 ÷ 4 = 24
80 ÷ 4 = 20	44 ÷ 2 = 22	75 ÷ 5 = 15
84 ÷ 4 = 21	51 ÷ 3 = 17	72 ÷ 6 = 12
88 ÷ 8 = 11	57 ÷ 3 = 19	78 ÷ 6 = 13
96 ÷ 8 = 12	60 ÷ 3 = 20	90 ÷ 6 = 15
51 ÷ 3 = 17	80 ÷ 4 = 20	
46 ÷ 2 = 23	60 ÷ 5 = 12	

EXERCISE II

Dividing three-digit numbers is very similar to dividing double-digit ones. Split the number you divide into "friendly" chunks, divide them separately, then add the results.

Problem: 108 ÷ 4 = ?

Solution: a) 108 is equal to 100 + 8.
Next 100 ÷ 4 = 25 and 8 ÷ 4 = 2. Then, 25 + 2 = 27.

b) 108 is also equal to 80 + 28. Next, 80 ÷ 4 = 20 and 28 ÷ 4 = 7. Then, 20 + 7 = 27.

108 ÷ 4 = 27

Problem: $126 \div 7 = ?$

Solutions: a) 126 is equal to 70 + 56. Next, $70 \div 7 = 10$ and $56 \div 7 = 8$. Then, $10 + 8 = 18$.

b) 126 is also equal to 63 + 63. Next, $63 \div 7 = 9$. Then, $9 + 9 = 18$.

$$126 \div 7 = 18$$

$80 \div 5 = 16$	$114 \div 6 = 19$	$135 \div 9 = 15$
$80 \div 4 = 20$	$112 \div 8 = 14$	$128 \div 8 = 16$
$100 \div 4 = 25$	$119 \div 7 = 17$	$153 \div 9 = 17$
$104 \div 4 = 26$	$120 \div 8 = 15$	$136 \div 8 = 17$
$110 \div 5 = 22$	$117 \div 9 = 13$	$144 \div 8 = 18$
$112 \div 7 = 16$	$120 \div 6 = 20$	$152 \div 8 = 19$
$108 \div 9 = 12$	$133 \div 7 = 19$	$144 \div 9 = 16$

SUM VERSUS PRODUCT

To help dividing double-and triple-digit numbers, in Verbal Math we look at the numbers as sum of two or more parts. Then, we divide each part separately and add the results (quotients).

We can also see the number we divide as the product of two factors. Remember that the numbers that we multiply to each other are called factors. We divide only one of two number and then multiply the quotient by the second factor. Let me show you.

Problem: $120 \div 3 = ?$
Solution: 120 is equal to 30×4.
Next, $30 \div 3 = 10$ (we divide one factor, not both).
Then, $10 \times 4 = 40$.

$$120 \div 3 = 40$$

Problem: $140 \div 7 = ?$
Solution: 140 is equal to 14×10. Next, $14 \div 7 = 2$.
Then, $2 \times 10 = 20$.

$$140 \div 7 = 20$$

WORD PROBLEMS

1. Olympic games are played every four years.
   a) How many games will be played in 36 years? *Ans:* 9 games.
   b) How many games are played in 48 years? *Ans:* 12 games.
   c) How many games are played in 60 years? *Ans:* 15 games.

2. A family divided 57 cows and 48 pigs equally among three children. How many cows and how many pigs did each get? *Ans:* 19 cows and 16 pigs.

3. A 160 page manuscript was divided equally among 4 proofreaders. How many pages did each receive? *Ans:* 40 pages (160 is equal to 16 × 10, then 16 ÷ 4 = 4 and 4 × 10 = 40).

4. Mr. Fullhouse took a deck of 52 cards and divided them in 4 equal stacks. How many cards are in each stack? *Ans:* 13 cards.

5. Then, he took out one card and divided the deck into 3 stacks. How many cards are in each stack now? *Ans:* 17 cards.

6. Then, he put two new decks of cards together and divided all the cards into 8 stacks. How many cards are in each stack? *Ans:* 13 cards (52 + 52 = 104, 104 is equal to 80 + 24. Next, 80 ÷ 8 = 10 and 24 ÷ 8 = 3. Finally 10 + 3 = 13). Also, if Mr. Fullhouse divided one deck into 4 parts and got 13 in each stack, then 2 decks divided into 8 stacks will also have 13 cards in each of stack.

7. Jason took 96 dragon teethes and divided them equally into 8 handfuls. How many teeth are in each handful? *Ans:* 12 dragon teethes. Have you read the Greek Myth about Jason and the Golden Fleece?

8. There were 46 chromosomes on a photograph and a researcher circled them into pairs. How many pairs did she circle? *Ans:* 23 pairs.
   Solution: Grouping chromosomes in pairs is the same as dividing them into groups of 2. Then 46 ÷ 2 is equal to (40 ÷ 2) + (6 ÷ 2) = 20 + 3 = 23.

9. How many legs do 44 panthers have? ***Ans:*** 176 legs.

10. How many panthers have 76 legs? ***Ans:*** 19 panthers.

11. If 78 piglets were born to 6 mama pigs. How many babies for each mama pig? ***Ans:*** 13 piglets.

12. If each poster board can fit 6 typed pages, how many boards does Nadia need for 72 pages? ***Ans:*** 12 boards (76 is equal to 60 + 12).

13. If a host puts 3 lumps of sugar in each cup, how many cups will use 48 lumps? ***Ans:*** 16 cups.

14. If a camp allows 4 people on each campsite, how many campsites will 64 campers fill? ***Ans:*** 16 sites.

15. If 108 pencils were divided into 6 boxes, how many are in each box? ***Ans:*** 18 pencils.

16. If one pair of glasses uses 4 screws, how many screws are needed for 44 pairs? ***Ans:*** 176 screws.

17. If 105 golf clubs were divided into 7 per each golf bag, how many bags are there? ***Ans:*** 15 bags.

18. On a field trip, 78 students sat in 6 rows to watch a modern dance show. How many sat in each row? ***Ans:*** 13 students.

19. With 6 peas in each pod there were 96 peas altogether. How many pods were there? ***Ans:*** 16 pods.

20. There are 5 fingers on each hand (we will count thumbs as fingers, even if funny looking). How many hands do 80 fingers make? ***Ans:*** 16 hands.
    a) How many 8-finger hands would 80 fingers make? ***Ans:*** 10 hands.
    b) How many 4-finger hands would 80 fingers make? ***Ans:*** 20 hands.
    c) How many hands with 7 fingers do 91 fingers make? ***Ans:*** 13 hands. Silly!

21. If 8 leaves are on each twig, how many twigs hold 96 leaves? ***Ans:*** 12 twigs.

22. If it takes 7 ants to carry a twig, how many twigs can 84 ants carry? *Ans:* 12 twigs.

23. If each ant has 6 legs, how many ants have 72 legs?
Ans: 12 ants.

24. If one glass holds 6 ounces, how many glasses could hold 84 ounces? *Ans:* 14 glasses.

25. Little Manush can carry 5 pounds of rice each trip from the market. How many trips Little Manush must take to bring home 65 pounds of rice? *Ans:* 13 trips.

26. If a CD player can hold 6 CD's, how many players will hold 72 CD's? *Ans:* 12 players.

27. How many garages are needed for 45 cars, if each garage can hold 3 cars? *Ans:* 15 garages.

28. A magazine page can fit either 4 long ads, or 6 medium size ads, or 8 short ads.
a) How many pages will it take for 48 long ads? *Ans:* 12 pages.
b) How many pages will it take for 48 medium size ads?
Ans: 8 pages.
c) How many pages will it take for 48 short ads? *Ans:* 6 pages.

29. A CD can hold either 7 short songs or 3 long ballads.
a) How many CD's will it take for 84 songs? *Ans:* 12 CD's.
b) How many CD's will hold 84 ballads? *Ans:* 28 CD's.

30. If it takes feather from 4 geese to make a pillow, how many pillows can 76 geese make?
Ans: 19 pillows (although, geese don't make pillows, people do).

31. A box can hold either 4 large cookies or 8 candies.
a) How many boxes will we need for 84 cookies? *Ans:* 21 boxes.
b) How many boxes will we need for 96 candies? *Ans:* 12 boxes.

32. An apple slicer cuts an apple into 6 pieces. How many apples do we need to make 96 pieces? *Ans:* 16 apples.

33. How many six string guitars together have 120 strings?
    Ans: 20 guitars.
    a) How many guitars have 126 strings? *Ans:* 21 guitars.
    b) How many guitars have 600 strings? *Ans:* 100 guitars.
    c) How many guitars have 300 strings? *Ans:* 50 guitars.
    d) How many guitars have 180 strings? *Ans:* 30 guitars.

34. If a puppet has 6 attached strings, how many puppets have 114 strings? *Ans:* 19 puppets.

35. If one dress has 7 buttons in the back, how many dresses have 105 buttons? *Ans:* 15 dresses.

36. A patrol car has 5 flashing lights on the top. How many patrol cars will have 115 lights? *Ans:* 23 cars.

37. A doctor told Tommy Ache to take 6 pills a day and wrote a prescription for 84 pills. How many days will Tommy take his medicine? *Ans:* 14 days.

38. She also told Tommy to put 2 drops of medicine in each ear twice a day for 5 days. How many drops does it make?
    Ans: 40 drops.
    Solution: 2 drops × 2 ears = 4 drops (each time). Next, 4 drops × 2 (times a day) = 8 drops (each day). Then, 8 drops (each day) × 5 (days) = 40 drops altogether.

39. During an intermission the theater buffet was selling 9 sodas every minute. How long did the intermission last if they sold 117 sodas? *Ans:* 13 minutes.

40. During the food drive student was asked to bring 5 items each. By the end of the drive the class collected 135 items. How many students participated?
    Ans: 27 students (135 is equal 50 + 50 + 35).

41. If a dragonfly has 4 wings how many wings do 64 dragonflies have? *Ans:* 256 wings.

42. How many octopuses have 128 tentacles? *Ans:* 16 octopuses or 16 octopi, or 16 sea creatures with eight arms and a mouth in the middle.

43. The price for one picture frame is $8. How many picture frames can you buy for $136? *Ans:* 17 frames (136 is equal 80 + 56).

44. Bahar spent $72 to buy $8 T-shirts. How many did she buy? *Ans:* 9 shirts.

45. If Bahar spent $135 to buy $9 shorts, how many did she buy? *Ans:* 15 shorts.

46. A master made 112 widgets and the trainee made 8 times fewer. How many did the trainee make? *Ans:* 14 widgets.

47. An expert sharpened 119 knives and scissors, a novice did 7 times fewer. How many knives and scissors did the novice sharpen? *Ans:* 17 knives and scissors.

48. A toolsmith made 153 doohickeys and divided them into 9 doohickey holders. How many are in each holder? *Ans:* 17 doohickeys. Even if you don't know what a doohickey is, you can still come up with the right answer.

49. Mrs. Gill and Mr. Fin spent $171 to buy 9 sharks. How much did they pay for each shark? *Ans:* $19.

50. When 108 shovels were divided among 6 digging teams, how many shovels did each team get? *Ans:* 18 shovels.

21

DIVISION OF NUMBERS
WITH REMAINDER

Sometimes we divide a number that doesn't exactly break into equal parts.

For example, 8 divided by 2 is equal 4. That means that there are 2 equal parts, 4 in each part. However, if we divide nine by two, one is left out. We say then, that the answer to the problem $9 \div 2$ is 4 with 1 being a left-over or a "remainder".

This way, $5 \div 2$ is equal 2 with remainder 1.

$10 \div 2 = 5$, but $11 \div 2$ is equal 5 with remainder 1.

$12 \div 4 = 3$, but $13 \div 4$ is equal 3, remainder 1. And $14 \div 3 = 3$, remainder 2.

$20 \div 5 = 4$, but $23 \div 5$ is equal 4, remainder 3.

The times table is important for us to match the closest product of two numbers with the number given in the problem. For example, when we divide 17 by 4 the closest number divided by 4 is 16. Then, $17 \div 4 = 4$ and the remainder 1. If we divide 18 by 4, then it is $18 \div 4 = 4$

(remainder 2). Next, 19 ÷ 4 = 4 (3) and 20 ÷ 4 = 5, there is no remainder here since the remainder can't be equal or bigger that the divisor.

Remainders can be large but not larger than the divisor. For example,

- 35 ÷ 20 = 1, remainder 15
- 43 ÷ 4 = 10, remainder 3
- 49 ÷ 9 = 5, remainder 4
- 99 ÷ 10 = 9, remainder 9
- 199 ÷ 100 = 1, remainder 99

EXERCISE I

4 ÷ 2 = 2	12 ÷ 3 = 4	21 ÷ 8 = 2 (5)
5 ÷ 2 = 2 (1)	15 ÷ 4 = 3 (3)	28 ÷ 9 = 3 (1)
5 ÷ 3 = 1 (2)	16 ÷ 5 = 3 (1)	28 ÷ 7 = 4
7 ÷ 3 = 2 (1)	19 ÷ 5 = 3 (4)	29 ÷ 3 = 9 (2)
12 ÷ 5 = 2 (2)	21 ÷ 4 = 5 (1)	37 ÷ 5 = 7 (2)
10 ÷ 3 = 3 (1)	21 ÷ 5 = 4 (1)	44 ÷ 5 = 8 (4)
11 ÷ 3 = 3 (2)	21 ÷ 6 = 3 (3)	45 ÷ 4 = 11 (1)

WORD PROBLEMS

1. A school bought 60 plants, 8 for each classroom and the leftovers for the office. How many classrooms are there and how many plants went to the office?
    Ans: 7 classrooms, 4 plants to the office.
    Solution: The closest number to 60 divisible by 8 is 7, because 8 × 7 = 56. Then, what's left is 60 - 56 = 4 (flowers to the office).

2. Ricardo divided 77 tin soldiers into regiments, 9 soldiers in each. The remainders were the officers. How many regiments and the officers did he make?
    Ans: 8 regiments (9 × 8 = 72) and 5 officers.

3. Leo divided 95 pictures, 10 pictures in each album. How many pictures were in the last, incomplete album? *Ans:* 5 pictures.
    Solution: The number closest to 95 divisible by 10 is 90. Then, 95 - 90 = 5).

4. Alfonso took $80 to buy ping pong rackets for the team, at $9 per racket. How many rackets did he buy and what was the change? ***Ans:*** 8 rackets, $8 (72 is the closest number to 80, divisible by 9).

5. A writer divided 57 stories into several books, 11 stories in each book, but some stories didn't make into the books. How many didn't? ***Ans:*** 2 stories (55 is the closest number to 57, divisible by 11).

6. A plumber cut a 70 inch pipe into 12 inch segments and threw away the rest. How long was the 'rest'?
Ans: 10 inches (60 is the closest number to 70, divisible by 12).

7. On a farm 50 eggs were divided into half-a-dozen (6) eggs per box. How many eggs did not make the full box?
Ans: 2 eggs (50 eggs - 48 eggs = 2 eggs).

8. For a screening 53 actors and actresses were divided into 7 equal groups. How many are in each group and in the last, incomplete group?
Ans: 7 people in each group and 4 in the last, incomplete group.

9. There were 39 ice cubes in an ice bucket. Lilly filled 9 glasses with 4 cubes each but the last glass had fewer. How many?
Ans: 3 cubes.

10. Colton had 59 minutes for a math test. He divided his time, 6 minutes per 9 problem plus few extra minutes to go over the answers. How many extra minutes did he have? ***Ans:*** 5 extra minutes.

11. There were 82 days of vacation time. How many weeks plus days does it make? ***Ans:*** 11 weeks and 5 days.

12. After Rhiannon divided all her plums into 5 bags, 7 plums in each bag, there were 5 plums left out. How many plums were there altogether? ***Ans:*** 40 plums.
Solution: Rhiannon had some plums. She put in the bags 7 (plums) × 5 (bags) = 35 plums. Then we add 5 plums there were left out, 35 + 5 = 40 (plums).

13. A tailor cut a ribbon into 6 equal pieces, 6 inches each and had 3 inches of ribbon left. How long was the ribbon?
    Ans: 39 inches (6 × 6 = 36; 36 + 3 = 39).

14. A farm has 12 pigs and a rooster. How many legs do all the animals have? *Ans:* 50 legs (4 × 12 = 48; then, 48 + 2 = 50).

15. There are 6 regular classes, 20 children in each class, and also a Special Day Class with 12 children. How many children are in the school? *Ans:* 132 children.

16. A hen laid 6 regular eggs and one small egg. All eggs together weigh 20 ounces. What's the weight of the small egg?
    Ans: 2 ounces.
    Solution: First we find the number closest but less than 20 divisible by 6. That is 18. That means, although the problem doesn't ask, that each regular egg weighs 3 ounces, 18 (ounces) ÷ 6 (eggs) = 3 ounces. That leaves 20 ounces - 18 ounces = 2 ounces for the small egg.

17. Robin Hood divided arrows into 3 quivers, 9 arrows in each quiver, and put remaining 5 arrows in the fourth. How many arrows does he have? *Ans:* 32 arrows.

18. Let's reverse the problem. Robin divided 32 arrows into quivers, 9 arrows in each quiver but the last quiver was incomplete. How many quivers did he use and how many arrows were in the last quiver? *Ans:* 4 quivers (3 complete, 1 incomplete), 5 arrows in the last quiver.
    Solution: The closest number less than 32 but dividable by 9 is 27. That means that first 27 arrows were placed in 3 quivers. Then we have 32 - 27 = 5 arrows left. They went to the incomplete quiver number 4.

19. Seven mount climbers divided all their gear into 8 items per person. There were still 4 items left. How many pieces of gear do they have? *Ans:* 60 pieces (8 × 7 = 56; 56 + 4 = 60).

20. There is a $10 bill on the table. The rest of the money is divided into 5 pockets, $8 in each. How much money is there altogether?
    Ans: $50.

21. For a paintball game 9 children divided into 4 teams. How many children were in each team and how many children were left out? *Ans:* 2 children per team, 1 child was left out.

22. Louis divided 15 shoes into pairs. How many pairs were there and how many shoes were without a pair? *Ans:* 7 pairs and 1 shoe with no pair.

23. John and Paul divided 25 gloves into pairs. How many pairs were there and how many gloves were without a pair?
Ans: 12 pairs and 1 lonely glove. All you need is [a] glove, glove. Glove is all you need... to make a pair.

24. Fatima divided 25 peaches to make 3-peach tarts. How many tarts she made and how many peaches were left out?
Ans: 8 tarts and 1 peach will remain.
Solution: The closest number less than 25 but divisible by 3 is 24. Then, 24 (peaches) ÷ 3 (peaches in each tart) = 8 (tarts) and there will 25 - 24 = 1 peach left.

25. Using 13 buttons a tailor made shirts, 5 buttons on each shirt and also one pair of shorts with left-over buttons. How many shirts did he make and how many buttons were on the shorts?
Ans: 2 shirts and 3 buttons for the shorts.

26. Ishmael caught 26 sardines and he threw out 2 sardines in the water to have them evenly into plastic bags. If he had 4 bags, how many sardines in each bag?
Ans: 6 sardines in each bag (26 - 2 = 24; 24 ÷ 4 = 6).

27. Mr. Rooster tried to divide 17 chickens evenly into 5 pens. How many chickens were in each pen and how many were left out?
Ans: 3 chickens in each pen, 2 chickens were left out and had to share Mr. Rooster's family room.

28. After filling four 8 oz glasses with milk, there was 5 oz of milk left in the pitcher. How many ounces of milk were in the pitcher at first? *Ans:* 37 oz.
Solution: Four glasses with 8 oz of milk is 8 (oz) × 4 = 32 (oz). Then, 32 oz + 5 oz (in the pitcher) = 37 oz.

29. A troop of girl scouts divided into 5 teams, 5 girls in each team. There were 4 girls left out. How big is the troop?
Ans: 29 girl scouts.

30. Delmar bought 50 nails to build birdhouses. If it takes 8 nails to build one birdhouse, how many houses did he build and how many nails were unused? *Ans:* 6 houses, 2 nails unused.
Solution: 48 is the closest number to 50 nails. If it takes 8 nails for each house, then Delmar can only make 6 houses. He can't have more groups of 8 nails beyond that. Delmar will make 6 houses and have 2 nails in remainder.

31. Liana was asked to buy milk with $20. The price of milk bottle is $3. How many bottles did she buy and how much was the change? *Ans:* 6 bottles and $2.

32. Tracy has 45 strings. If it takes 6 string for each guitar, how many can she string and how many strings will remain?
Ans: 7 guitars, 3 extra strings.

33. A 20 inch candle burns 2" every hour. How many hours did the candle burn if at the end it was 14"? *Ans:* 3 hours.
Solution: Altogether the candle burned 20 - 14 = 6 inches. If it burns 2" every hour, then 6 ÷ 2 = 3 (hours).

Quartet is a group of four musicians. There were 51 musicians waiting for quartet audition. How many quartets were there and how many musicians were not making a quartet?
Ans: 12 quartets and 3 musicians. Did you know that one musician or singer is a soloist, two make a duet, three is trio, five – quintet, six – sextet, seven – septet, and eight – octet, nine – nonet. After that, I guess, it's just an orchestra.

34. If the distance is 27 miles and Ms. Walker makes 4 miles every hour, how many complete hours did she walk and how many miles were in the last part of her journey?
Ans: 7 hours and 3 miles. Part of a journey is called a leg, even though it would take two legs to make it.

35. A factory made 46 cuckoo clocks and mailed them 4 in a box. How many boxes they mailed and how many clocks were in the last, incomplete box?
   Ans: 11 boxes and 2 clocks in the last box. I wonder how much surprised will be the cuckoo if we'll face it against the wall?

36. In a small theater 67 seats are divided into 9 rows, 8 rows with the same number of seats and the last in the back with a few remaining seats. How many seats are in each row including the last?
   Ans: 8 rows with 8 seats each and 4 seats in the back.

37. Faustus gets $15 weekly allowance that he spends every morning on a cup of hot chocolate. How much does his drink cost if at the end of the week he has $1 saved.
   Ans: $2 per a cup of hot chocolate.
   Solution: If Faustus gets $15 and has $1 left, then for 7 days of the week he has $14 or $2 per day.

38. In the spring 74 birds flew into town. Of those, 23 sat on trees and the rest evenly divided among 3 electric wires. How many birds sat on each wire?
   Ans: 17 birds (74 - 23 = 51. Then, 51 ÷ 3 = 17). Small birds seating on electric wires don't get hurt. According to physics laws, electricity goes through them without causing harm.

39. A contractor built 73-unit apartment building, 9 units on each floor, except for the top floor. How many floors are in the building and how many apartments are on the top floor?
   Ans: 8 floors, 1 apartment called penthouse.

40. All in all, 72 ounces of honey were divided into 8 oz jars. How many full jars were filled and how much was in the last jar?
   Ans: 9 full jars, the last jar also had 8 oz, because 72 divides by 8 without remainder.

41. In a safari park, 54 visitors completely filled 7 passenger vans with the remaining 5 in a jeep. How many were in each van?
   Ans: 7 visitors (49 is the closest number to 54, divisible by 7).

42. In the zoo 58 parrots were divided 8 parrots per cage and the rest were let free. How many cages were filled and how many parrots were let free?
Ans: 7 cages and 2 parrots went free.

43. A flower shop put 83 roses into bunches of 7. How many bunches did they make and how many roses did not make into a bunch?
Ans: 11 bunches plus 6 roses.

44. Mr. Chavez saw 49 grapes on a plate and began sticking them into his mouth, 6-grape fist at a time (he was hungry). How many fists of grapes did it take to finish the plate and how many grapes were in the last fist?
Ans: 9 fists of grapes altogether, 8 fists with 6 grapes and 1 fist with only one grape.

45. Grandma picked 70 pickles for pickling, filling each jar with 8 pickles. She put the remaining pickles into salad. How many jars did she make and how many pickles went into salad?
Ans: 8 jars (each with 8 grapes) and 6 pickles into salad.

46. Moving to another town Lionel put his collection of 93 train cars into boxes, 7 cars in each box. How many boxes did he use, and how many trains went into the last, unfilled box?
Ans: 13 boxes and 2 trains in the last box.

47. In a store 84 pounds of oranges were split into 4 boxes and each box was further divided evenly into 7 bags. How many pounds of oranges were in each bag?
Ans: 3 pounds ($84 \div 4 = 21$, $21 \div 7 = 3$).

48. Olaf brought home gold fish, 6 plastic bags, 9 fish in each bag. He put all fish together and then divided the fish into 3 tanks. How many fish are in each tank?
Ans: 18 fish ($6 \times 9 = 54$, $54 \div 3 = 18$)

49. A washing machine took 12 items of clothes in each of 6 loads. Then, all washed clothes were divided into 9 runs for the dryer. How many items were in each dryer run?
Ans: 8 items ($12 \times 6 = 72$, $72 \div 9 = 8$).

DIVISION AND INTRODUCTION TO FRACTIONS

We know already that one-half of a number is a number divided by two. One-third is the number divided by three, and one-quarter is the number divided by four. There are of course one-fifth, or a number divided by five; one-sixth, or divided by six; and so on.

When we say we multiply a number by one-half that means we are dividing it by 2.

What is one-half of 4? The question asks what is one half of four or, in other words, what are two equal parts when put together will make 4? The answer is 2, because 4 ÷ 2 = 2.

What is one-third of 12? The question is asking which 3 equal parts together make 12. In other words, what is 12 ÷ 3? ***Ans:*** 4

- One-half of something means divide number by 2
- One-third of something means divide number by 3
- One-fourth of something means divide number by 4
- One-fifth of something means divide number by 5
- One-sixth of something means divide number by 6

- One-seventh of something means divide number by 7
- One-eighth of something means divide number by 8
- One-ninth of something means divide number by 9
- One-tenth of something means divide number by 10

EXERCISE I

1. What is one-half of 10? *Ans:* 5

2. What does one-eleventh mean? *Ans:* divide number by 11.

3. What is one-half of 30? *Ans:* 15

4. What is one-half of 44? *Ans:* 22

5. What is one-half of 100? *Ans:* 50

6. What is one-half of 0? *Ans:* 0, of course, because half of nothing is nothing.

7. Now, what is one and a half times the number? That means that we add one-half of a number to the number.

8. What is one and a half of 8? *Ans:* 12, because 8 + 4 (one-half of 8) = 12

9. What is one and a half of 18? *Ans:* 27, because 18 + 9 (one-half of 18) = 27

10. What is one and a half of 30? *Ans:* 45, because 30 + 15 (one-half of 30) = 45

11. What's bigger one half or one-quarter of a number?
    Can you explain and give examples?
    Ans: one-half is bigger because when we divide a number into 2 parts, each part (quotient) is bigger than if we divide it by 4. Examples: 16 ÷ 2 = 8, but 16 ÷ 4 = 4, and 8 is bigger than 4.

12. What is one-half of 18? *Ans:* 9

13. What is one-quarter of 16? *Ans:* 4

14. What is one seventh of 63? *Ans:* 9

15. What is smaller one-forth or one-eighth of a number? **Ans:** one-eighth, because when you keep dividing a number into more parts, each part will get smaller. Try this with pizza.

   - One-half of 2 is 1
   - One-third of 3 is also 1, because $3 \div 3 = 1$
   - One-tenth of 10 is 1, as well.

16. What is one-fifth of 5? **Ans:** 1

17. What is one-quarter of 4? **Ans:** 1

18. What is one-seventh of 7? **Ans:** 1

19. What is one-ninth of 9? **Ans:** 1

20. What is one-hundredth of 100? **Ans:** 1

EXERCISE II

1. How many times 72 is more than 9? **Ans:** 8 times.

2. How many times 60 is more than 10? **Ans:** 6 times.

3. How many one-fives are in 20? **Ans:** 4 one-fives.

4. Or you can ask "What is one-fifths of 20?" **Ans:** 4

5. How many one-fours are in 20? **Ans:** 5 one-fourths

6. Or you can ask "What is one-forth of 20?" **Ans:** 5

7. How many one-sevens are in 42? **Ans:** 6, in other words, what is $42 \div 7$?

8. What is one-seventh of 42? **Ans:** 6

9. What is one-sixth of 42? **Ans:** 7

10. How many one-eights are in 32? **Ans:** 4, in other words, what is $32 \div 8$?

11. What is one-eighth of 32? **Ans:** 4

12. What is one-fourth of 32? **Ans:** 8

13. How many one-fifths are in 45? **Ans:** 9

14. What is one-fifth of 45? **Ans:** 9

15. What is one-ninth of 45? **Ans:** 5

16. How many one-fourths are in 24? *Ans:* 6

17. How many one-sixths are in 24? *Ans:* 4

18. How many one-tenths are in 50? *Ans:* 5

19. How many one-fifths are in 50? *Ans:* 10

20. What is one-quarter of 28?
    Ans: 7, in other words, what is 28 ÷ 7?

21. What is one-eighth of 72? *Ans:* 9

22. What is one-tenths of 100? *Ans:* 100

23. What is one-half of 24? *Ans:* 12

24. What is one-ninth of 63? *Ans:* 7

25. What is one-seventh of 56? *Ans:* 8

WORD PROBLEMS

1. Neil earned $12 and Nelly earned half as much. How much?
   Ans: $6

2. There are 12 people in one house and one-third of that number in the shed. How many people are in the shed? *Ans:* 4 people.

3. I pour 15 gallons of milk in one can but only one-third of that amount into the other. How much milk does the smaller can hold? *Ans:* 5 gallons.

4. There are dozen eggs in the basket.
   a) If I took out one-half, how many were left? *Ans:* 6 eggs.
   b) Bob took out one-third, how many? *Ans:* 4 eggs.
   c) Cathy took out one-quarter, how many? *Ans:* 3 eggs.
   d) Don took out one-sixth, how many? *Ans:* 2 eggs.
   e) Eve took out one-twelfth, how many? *Ans:* 1 eggs.

5. Hank puts 8 spoonfuls of sugar in his tea, Hugh puts one-quarter of that. How many spoons of sugar are in Hugh's tea?
   Ans: 2 spoons.

6. Lena ate 32 candies and Louis had one-eighth of that. How many candies did Louis have? *Ans:* 4 candies. You might say that eating 32 candies is bad for Lena's teeth, but Lena is 93 and already lost all her teeth.

7. Sandra asked 36 famous writers for the autographs and only one-quarter of them replied. How many writers wrote back? **Ans:** 9 writers.

8. One shop has 42 ice-cream flavors and pop sickles flavors one-sixth of these. How many pop sickle flavors do they have? **Ans:** 7 flavors.

9. There are 42 students in the ballet class, one-seventh are boys. How many girls are there?
    Ans: 36 girls (there are 6 boys, then 42 - 6 = 36 girls).

10. Out of 56 photos in Ms. Cricket's album, one-seventh are insects. How many insects pictures are there? **Ans:** 8 pictures.

11. There are 72 patients in the emergency room, one-eighth are children. How many children are in the ER? **Ans:** 9 children.
    a) One-third of the children in the ER have fever. How many? **Ans:** 3 children.

12. Judge Taylor bought 54 notebooks, one-ninth of them were legal pads. How many legal pads did he buy?
    Ans: 6 legal pads. The difference between the legal and regular (also known as letter) pads is page length÷ legal page is 14 inches long, letter page is shorter, only 11 inches, but don't call it illegal.

13. Last year Jeremy Germ took 48 sick days, this year he took only one-sixth of them. How many sick days did he take this year? **Ans:** 8 sick days.

14. Together, all our courtyard kittens have 32 paws. In the neighbor's courtyard they have one-eighth of the paws. How many paws and kittens do the neighbors have? **Ans:** 4 paws or one kitten.

15. There are 45 stamps on the package and one-fifth of that on the letter. How many stamps are on the letter? **Ans:** 9 stamps.

16. On the farm one-eighth of 56 workers drove tractors. How many workers didn't? **Ans:** 49 workers (one-eighth of 56 is 7 workers; then 56 - 7 = 49).

17. From 27 crayons on the teacher's desk each student took 4 until only a few crayons were left on the desk. How many students were in class and how many crayons were left on the desk? *Ans:* 6 students and 3 extra crayons (24 is the closest number to 27, divisible by 4).

18. The seeds of lotus flower can last 100 years, an orchid seed lasts 2 years. How many times longer can lotus seed last? *Ans:* 50 times longer.

19. A dry sponge weighs 7 oz, the wet sponge is 63 oz. How many times a wet sponge is heavier than the dry and what's the weight of the liquid? *Ans:* 9 times, 56 oz of liquid (63 oz - 7 oz = 56 oz).

20. At rest a whale's heart rate was 9 beats per minute, when racing it was 72 beats per minute. How many times whale's resting heart rate is slower than racing?
Ans: 8 times. To answer the question 'how many times' we need to divide the racing heart rate (72 beats per minute) by the resting rate (9).

21. After 36 peaches were divided evenly among 8 children, some were left in the box. How many peaches did each child get and how many were left in the box?
Ans: 4 peaches for each child and 4 peaches left in the box (the closest number to 36 divisible by 8 is 32).

22. In the cabin 66 wood logs were divided into 9 equal piles. How many logs were in each pile and how many were left out?
Ans: 7 logs in each pile, 3 left out.

23. After 79 CD's were evenly divided on 7 shelves. How many CD's were left out? *Ans:* 2 CD's, (77 CD's divided into 11 CD's for each shelf, and 2 CD's left out).

24. Out of 50 muffins, one-fifth were with raisins. How many were not? *Ans:* 40 muffins with no raisins.
Solution: One-fifth of 50 is 10 (muffins with raisins). Then, 50 (all muffins) - 10 (with raisins) = 40 muffins without raisins.

25. What is one-half of 20? *Ans:* 10

26. What is one-half of 40? *Ans:* 20

27. What is one-half of 60? *Ans:* 30

28. What is one-quarter of 20? *Ans:* 5

29. What is one-quarter of 40? *Ans:* 10

30. What is one-quarter of 80? *Ans:* 20

31. A giant watermelon weighs 60 pounds. After one-quarter was cut out, how much did the rest weigh?
    Ans: 45 pounds. By the way, the World's Largest Watermelon weighed 268.8 pounds!

32. A race pigeon flies 48 miles in one hour.
    a) How many miles will it fly in one-quarter of an hour?
    Ans: 12 miles.
    b) How many miles will it fly in one -sixth of an hour?
    Ans: 8 miles.
    c) How many miles will it fly in one-eighth of an hour?
    Ans: 6 miles.

33. How many hours are in one-eighth of a day?
    Ans: 3 hours, because the day is 24 hours.

34. At the World Climate Congress geologists made one-quarter of 64 scientists. How many? *Ans:* 16 geologists.
    a) One-eighth were biologists. How many? *Ans:* 8 biologists.
    b) One-half were astronomers. How many?
    Ans: 32 astronomers.

35. There are 30 students in the class. One-fifth are top students. How many students are not in the top group? *Ans:* 24 students (one-fifth of 30 is 6, 30 - 6 = 24).

36. Six zookeepers cleaned one-eighth of 96 cages. How many did they clean? *Ans:* 12 cages.

37. Out of 63 chewing gum sticks one-ninth are watermelon flavor. How many? *Ans:* 7 sticks.

38. Out of 72 volcanoes on a planet, one-eighth are active. How many? *Ans:* 9 volcanoes.

39. If out of 52 cards in the deck, one-quarter are hearts, how many hearts cards are in the deck? *Ans:* 13 cards.

40. There are 100 M&M's in the bag, one tenth are blue. How many? **Ans:** 10 M&M's.

41. Out of 20 animals in a zoo, one-twentieth are the elephants. How many elephants are the zoo? **Ans:** 1 elephant.

42. If out of 80 children in fourth grade only one-tenth received the flu shots, how many didn't? **Ans:** 72 children.
    Solution: One-tenth of 80 is 8 children (who received the shots). Then, 80 - 8 = 72 (children who didn't).

43. An egg weighs 60 grams.
    a) If one tenth of the egg's weight is its shell, how much does the shell weigh? **Ans:** 6 grams.
    b) One-half of the egg's weight is the white part called glair. How much does that weigh? **Ans:** 30 grams.
    c) The rest of the egg is yolk. How much does the yolk weigh? **Ans:** 24 grams, because 60 (the whole egg) - 6 (the shell) - 30 (the glair) = 24 (grams).

44. In a 90-head herd, one-tenth are goats, one-ninth are sheep, and the rest are cows. How many cows are in the herd? **Ans:** 71 cows (out of 90 animals there are 9 goats and 10 sheep, then 90 - 9 - 10 = 71).

45. In a 42-student class, one-sixth of the names start with the letter "J". How many students have their names starting with letters other than "J"? **Ans:** 35 students.

46. One-quarter of 52 kids living on our street Exercise more than half an hour a day. How many Exercise less than that? **Ans:** 39 kids.

47. A team played 35 games and lost one-fifth of them. How many games did they win or tied? **Ans:** 28 games.

48. In a "kids of steel" competition, the participants cover 20 kilometers. They swim one-twentieth and run one-quarter of the way. The rest they bike. How many kilometers do they bike? **Ans:** 14 kilometers, because 20 km - 1 km (swim) - 5 km (run) = 14 km.

49. If an average adult sleeps one-third of the day, how many hours is that? **Ans:** 8 hours.

50. Out of $40 Jamal spent one-eighth on pencils and one-fifth on the erasers. How much money were left for the books?
Ans: $27, because $40 - $5 - $8 = $27.

51. Two cars, 72 miles apart, are moving toward each other. After a while, the first car drove one-fourth and the second one-eighth of the way. How far did each car go and what is the distance between them?
Ans: First car drove 18 miles, the second did 9 miles, and now there are 45 miles between them.
Solution: First car drove 72 ÷ 4 = 18 (miles), the second car 72 ÷ 8 = 9 miles. Then, 72 - 18 - 9 = 45 (miles).

52. Mr. Lem saw 45 aliens in his dream, one-fifth of them were Martians. How many Martians were in Mr. Lem's dream?
Ans: 9 Martians.

53. Out of 56 grapes in a fruit vase one-eighth are seedless. How many grapes have seeds? **Ans:** 49 grapes.

54. How many weeks are in a 31 day months and how many days don't make the whole week? **Ans:** 4 weeks and 3 days.

55. How many months have 28 days? **Ans:** All of them.

23

FRACTION CONCEPTS

A whole, spells w-h-o-l-e, is the entire or complete number, or all of it. It's different from a hole, which spells h-o-l-e, which is an empty space like the one in the middle of a doughnut.

If we take a bunch of pencils and divide them into two equal sets, then each set is one-half of the whole bunch. If we divide the bunch into three equal sets, then each set is one-third. Dividing into 5 sets gives us fifths, divided into seven sets will have one-seventh of the whole, and so on.

If we take a box of nails and equally divide them into two piles, each pile will have one-half of the nails, but two one-halves together will make whole. A whole is also 3 one-thirds, 4 one-quarters, 5 one-fifths, and 10 one-tenths. When you break a number into smaller equal parts and then bring *all* these parts together, they'll make the whole number again.

What is bigger two-halves or three-thirds?
Ans: They are the same. Both make a whole.

If a whole is made of two halves, then one half plus another half make whole. If the whole is three-thirds, then two-thirds and one third together will make one whole.

What is bigger one-quarter or one-third of 12?
Ans: One third, of course. Because one-third of 12 is 4, but one-quarter of 12 is only 3.

What's bigger one-half or one-tenth of 20?
Ans: one-half.

What's bigger one-quarter or one-sixth of a number?
Ans: one-quarter, because if you take the same number and break it into more parts, each part will be smaller.

Think about it this way: You have the whole $30 and you divide it between either four friends or six, so each one gets either one-quarter or one-sixth. Then, one-quarter of $30 (or $15) will be bigger amount than one-sixth ($5).

EXERCISE I

1. What's bigger one-fifth or one-sixth of 30? **Ans:** one fifth, because $6 is bigger than $5,

2. What's bigger one-quarter or one-third of 24?
   Ans: one third or 8.

3. What's bigger one-seventh or one-ninth of a number?
   Ans: one-seventh.

4. What's bigger one-eighth or one-fifth of a number?
   Ans: one-fifth.

5. What's bigger one-eighth or one-ninth of a number?
   Ans: one-eighth.

6. What's bigger one-tenth or one-eleventh of a number?
   Ans: one-tenth.

7. What's smaller one-seventh or one-sixth of a number?
   Ans: one-seventh.

8. What's smaller one-thirtieth or one-twentieth of a number?
   Ans: one-thirtieth.

9. What's smaller one-thirtieth or one-thirteenth of a number?
   Ans: one-thirtieth.

EXERCISE II

If one-third of a number is the number divided by 3, then two-thirds of that number is the number divided by three and the 2 one-thirds added together.

Problem: What is two-thirds of 18?
Solution: One-third of 18 is 6. Then, two-thirds is 12
$(6 + 6 = 12$, or $6 \times 2 = 12)$

Problem: What is three-quarters of 20?
Solution: One-quarter of 20 is 5. Then, three-quarters is 15
$(5 \times 3 = 15)$

Problem: What is two-fifths of 35?
Solution: One-fifth of 35 is 7. Then, two-fifths is 14 $(7 \times 2 = 14)$

1. What's one-half of 20? *Ans:* 10

2. What's one-tenth of 20? *Ans:* 2

3. What's two-tenths of 20? *Ans:* 4

4. What's two-thirds of 12? *Ans:* 8
   (one-third of 12 is 4, then two one-thirds are 8)

5. What is two-thirds of 27? *Ans:* 18
   (one-third of 27 is 9, the two one-thirds is 18)

6. What's two-fifths of 20? *Ans:* 8 (because one-fifth is 4)

7. What's two-sevenths of 35? *Ans:* 10

8. What's two-thirds of 30? *Ans:* 20

9. What's two-tenths of 50? *Ans:* 10

10. What's three-tenths of 50? *Ans:* 15

11. What's two-ninths of 54? *Ans:* 12

12. What's four-ninths of 54? *Ans:* 24

13. What's three-quarters of 24? *Ans:* 18

14. What's three-fifths of 15? *Ans:* 9

15. What's three-quarters of 32? *Ans:* 24

16. What's three-fifths of 25? *Ans:* 15

17. What's four-fifths of 25? *Ans:* 20

EXERCISE III

One-half is twice as big as one-quarter. One half is a number divided by 2, while one-quarter is the number divided by 4. For example:

- One-half of 20 is 10.
- One-quarter of 20 is 5.
- 10 is twice a big as 5.

1. What's one-tenth of 30? *Ans:* 3

2. What's two-tenths of 30? *Ans:* 6

3. What's one-fifth of 30? *Ans:* 6

4. What's one-half of 16? *Ans:* 8

5. What is one-quarter of 16? *Ans:* 4

6. What is two-quarters of 16? *Ans:* 8
    Did you notice that one-half is equal to two-quarters?

7. What's one-tenth of 30? *Ans:* 3

8. What's two-tenths of 30? *Ans:* 6

9. What's one-fifth of 30? *Ans:* 6
    Did you notice that one-fifth is equal to two-tenths?

10. What's one-third of 24? *Ans:* 8

11. What's one-sixth of 24? *Ans:* 4

12. What's two-sixths of 24? *Ans:* 8
    Did you notice that one-third is equal to two-sixths?

13. What's one-tenth of 100? *Ans:* 10

14. What's seven-tenths of 100? *Ans:* 70

15. What's eight-tenths of 100? *Ans:* 80

16. What is one-half of 120? *Ans:* 60

17. What is one-quarter of 120? *Ans:* 30 because

18. What is one half of 200? *Ans:* 100

19. What is one quarter of 200? *Ans:* 50

20. What is two quarters of 200? *Ans:* 100

WORD PROBLEMS

1. There are 30 players on a school baseball team and two-thirds are freshmen. How many freshmen are there? *Ans:* 20 players.

2. It took 21 hours for the first person to swim across the English Channel in 1875. Now the record stands at one-third of this time. How many hours? *Ans:* 7 hours.
   The youngest person to swim across was Lynn Cox, she was 15. The oldest person to swim across, George Brunstad, was 70 years old at that time.

3. If I had $27 and spent two-thirds of my money today, how much did I spend? *Ans:* $18 (one-third of 27 is 9, then two-thirds is 18).

4. A crow ate eight-eighths of 48 grapes on a bunch, how many? *Ans:* All 48 of them.

5. An oracle made 50 predictions and only two-tenths of them came true. How many? *Ans:* 10 predictions.
   Ancient Greeks believed that the oracles can tell the future, answer hard questions, and give advice.

6. Hermosa finished two-thirds of a 45-page book. How many pages did she read? *Ans:* 30 pages.

7. Emma stored 15 pounds of dog food, two-thirds for the big dog and the rest for the puppy. How much is for the big dog?
   Ans: 10 lb.
   a) How much is for the puppy? *Ans:* 5 lb.

8. A baseball game lasts 9 innings. If we played two-thirds of the game, how many innings did we play? *Ans:* 6 innings.

9. Trudy worked three-quarters of the days in a 28 days month. How many days did she work? *Ans:* 21 days.
   Solution: One-quarter of a 28 days month is 7 days, three-quarters is 21 days.

10. Caleb worked two-thirds of the days in a 30 days month. How many days did he work? ***Ans:*** 20 days.

11. If there are 27 lessons in a book and a student finished four-ninths of them, how many lessons did she do? ***Ans:*** 12 lessons.
a) Another one has done two-ninths of the lessons, how many?
Ans: 6 lessons.
b) The third student has done five-ninths. How many lessons?
Ans: 15 lessons.

12. If it takes 28 minutes to get from train station to downtown, how long do three-quarters of the way take? ***Ans:*** 21 minutes.

13. If a dozen eggs cost 80¢, how much will you pay for three-quarters of a dozen? ***Ans:*** 60¢.
Solution: One quarter of 80¢ is 20¢, then three-quarters would be 60¢. Knowing that a dozen is 12 eggs doesn't matter for this problem.

14. What is four-fifths of 50? ***Ans:*** 40

15. What is four-tenths of 100? ***Ans:*** 40

16. What's bigger one-half or one-third? Hint÷ use number 12 to find out. ***Ans:*** One-half, because one-half of 12 is 6 and one-third is only 4.

17. Demitra drank seven-eighths of a 16 ounce bottle of water. How many ounces did she drink? ***Ans:*** 14 ounces.

18. Which one is bigger, two-thirds or three-quarters of 24?
Ans: Three-quarters is bigger, because two-thirds of 24 is 16, but three-quarters is 18.

19. Which one is bigger, two-thirds or three-fifths of 30?
Ans: Two-thirds is bigger, because two-thirds of 30 is 20, but three-fifths is only 18.

20. Jerry gave three-quarters of his baseball cards to his sister. If he had 48 cards, how many did he give away? ***Ans:*** 36 cards.

21. A magician showed only three-fifths out of 45 secret tricks to his apprentice (trainee). How many secretes did he teach?
Ans: 27 secrets.

22. Mr. Hand spent two-sevenths of $63 on a new watch. How much did the watch cost? *Ans:* $18.

23. What is bigger, three-quarters or three-fifths? Hint: use number 20 to figure out. It's easier with real numbers. *Ans:* Three-quarters.

24. What is bigger, four-seventh or four-fifths? A hint: use number 35 to figure out. *Ans:* Four-fifths.

25. There were 54 staples in a stapler and Hunter used five-ninths of them. How many staples did he use? *Ans:* 30 staples.

26. There were 42 cars on a parking lot. By noon we washed three-sevenths of them. How many cars did we wash? *Ans:* 18 cars.

27. One runny and sneezey day Miriam used three-quarters of her 32 handkerchiefs. How many did she use that day? *Ans:* 24 handkerchiefs. Achoo!

28. Only three-fifths of 35 hangers have cloths on them. How many? *Ans:* 21 hangers.

29. In a 28-flower bouquet, four-sevenths of all flowers are daisies. How many daisies are in the bouquet? *Ans:* 16 flowers.

30. There were 54 horses in a stable and Forrest took five-sixths of them out. How many horses went out? *Ans:* 45 horses.

31. Angelo runs 6 miles in 36 minutes. Brian runs it in five-sixths of that time. How fast is Brian? *Ans:* 30 minutes.

32. Playing softball with her brother Esther hit six-sevenths of 42 balls. How many hits was that? *Ans:* 36 hits.

33. In a 54-student chorus, two-ninths of the students can't sing. How many can't sing? *Ans:* 12 students.

34. In a 24-page college newspaper three-eighths are sports pages. How many sports pages are in the paper? *Ans:* 9 pages.

35. A catamaran on a 63 mile journey covered five-sevenths of the way the first week. How many miles was that? *Ans:* 45 miles.

36. If Monica spilled two-seventh of her 35 oz milk drink, how many ounces is that? *Ans:* 10 oz. Monica's cat invited neighbor cats and they had a party.

37. In one minute a chicken ate seven-ninths of 81 seeds. How many did it eat? *Ans:* 63 seeds.

38. There are 20 musicians and 20 chairs in an orchestra. If violin players grabbed one-quarter of all chairs, how many chairs were left for nonviolent non-violin players? *Ans:* 15 chairs.
Solution: Violin players snatched one-quarter of 20 chairs, or 5 chairs. Left were 20 - 5 = 15 chair for non-violin players.

39. If Armand spent one-sixths of a 24-hour day skateboarding, how many hours did he skate? *Ans:* 4 hours.

40. A demolition crew tore down eight-ninths of a 27 feet old wall. How many more feet are left for destruction? *Ans:* 3 feet.
Solution: First, we find out how much of the wall was torn down: eight-ninths of 27 is 24 (feet). Then, 27 - 24 = 3 (feet).

41. Maryanne used eight-ninths of her 90 stickers for a scrapbook. How many stickers are left? *Ans:* 10 stickers.

42. On a foggy day, Francois was able to see only one-third of the Eiffel Tower. What portion was hidden by the fog?
Ans: two-thirds, because the whole tower is three-thirds. Eiffel Tower, the tallest structure in Paris, France was built in 1889 and stands 1,063 ft tall.

43. During a commuter plane takeoff the door opened and two-sevenths of 56 suitcases fell out. How many?
Ans: 16 suitcases.

44. Reza spent five-ninths of his $81 savings on a new cell phone. What was the price of the phone? *Ans:* $45.

45. Lee has 63 items for laundry. If the first load was four-ninths of the items, how many items were there? *Ans:* 28 items.

46. A crew paved three-sevenths of a 77 feet driveway. How many feet are paved? *Ans:* 33 feet.

47. Goliath wears 45 size shoes. David wears shoes two-ninths that size. What's the size of David's shoes? *Ans:* size 10.

48. A very talkative duck quacks 96 times a minute. A quieter duck quacks two-eights of that. How many quacks is that?
Ans: 24 quacks.

49. A loud dog barked 40 times. A gentle dog barked two-eighth of that. How many times? **Ans:** 10 times.

50. Reba solved three-quarters and Simon solved two-thirds out of 24 problems on the test. Who solved more problems?
Ans: Reba solved more.
Solution: Three-quarters of 24 is 18 (the number of problems Reba solved). Two-thirds of 24 is 16 (Simon solved). Reba solved more.

24

ADDING AND SUBTRACTING FRACTIONS

Sometimes we can figure out a fraction without even knowing the exact number. If I say there were some apples on the table and Jill took one-half of them, then what's left is the other half.

Kim took one-third of the oranges on the table. Since there were three-thirds of the oranges and Kim took one-thirds of the whole, then there are two-thirds of the oranges left (three-thirds minus one-third equals two-thirds).

Now, someone took two-fifths of the pears from the table. How much of them are still on the table? Three-fifths, because the whole is five-fifths, from five-fifths we took away two-fifths and now three-fifths are left on the table. The exact number of pears on the table is not important for this problem, only the portion of the whole.

Then, if you know the fraction of pears left on the table (three-fifths) you can solve the problem for any number of pears: 15 (three-fifths of 15 is 9), 30 (three-fifths of 25 is 18), even 100 (three-fifths of 100 is 60).

EXERCISE I

1. Add one-half to one-half, how much is it? *Ans:* two-halves or the whole.

2. Add one-third to one-third, how much is it? *Ans:* two-thirds.

3. Add another one-third to the sum, what is it now?
   Ans: three-thirds or one whole.

4. What is two-fifths plus two-fifths? *Ans:* four-fifths.

5. What is four-ninths and two-ninths? *Ans:* six-ninths.

6. Add two-ninths to the sum, what is it now? *Ans:* eight-ninths.

7. What is three-elevenths plus seven-elevenths?
   Ans: ten-elevenths.

8. What is six-seventeenths plus five-seventeenths?
   Ans: eleven seventeenths.

9. What is one-quarter plus one-quarter? *Ans:* two-quarters or one-half (imagine 25¢ coins - the quarters).

EXERCISE II

1. Take one-quarter from the whole amount, how much is left?
   Ans: three-quarters, because the whole is made of four one quarters or four-quarters.

2. Take three-quarter from the whole, how much is left?
   Ans: one-quarter.

3. Take one-fifth from the whole, how much is left?
   Ans: four-fifths.

4. Take one-tenth from the whole, how much is left?
   Ans: nine-tenths.

5. Take two-fifths from the whole, how much is left?
   Ans: three-fifths.

6. Take three-sevenths from the whole, how much is left?
   Ans: four-sevenths.

7. Take five-ninths from the whole, how much is left?
   Ans: four-ninths.

8. Take four-tenths from the whole, how much is left?
   Ans: six-tenths.

9. Take nine-tenths from the whole, how much is left?
   Ans: one-tenth.

WORD PROBLEMS

1. If the whole is three-thirds and you took away one-third, how much is left? *Ans:* two one-thirds or two-thirds.

2. If I had two one-fifths and I bought three more, how many one-fifths do I have now? *Ans:* five one-fifths, five-fifths or a whole.

3. If Mary-Kate brought her five-tenths and Ashley carried her five-tenths, how much did both carry? *Ans:* ten-tenths or a whole.

4. If there were twenty-twentieths and a sneaky thief stole three-twentieths, how many twentieths were not stolen? *Ans:* seventeen-twentieths (20 - 3 = 17), still plenty of the twentieths to get by.

5. A bull went to a china shop and broke nine-tenths of all china. What portion was spared? *Ans:* one-tenth.

6. If you had fourteen-fourteenths and then learned that three of these fourteenths are not yours, how many fourteenths belong to you? *Ans:* eleven-fourteenths.

7. Eva did one-third of the team project and Fran also did one-third. What portion of the project did both do? *Ans:* two-thirds.

8. If one-quarter of the earthworm sticks out, what portion of the worm is inside the ground? *Ans:* three-quarters.
   a) Soon after, another one-quarter of the worm came out. What portion of the worm is out now? *Ans:* two-quarters or one-half.
   b) Then, one more quarter came out. How many quarters are still in underground? *Ans:* one quarter.

9. It took three-eighths of the day to fill two-fifths of a trash hauler. What portion of the hauler needs to be filled to make it full? *Ans:* three-fifths. Ignore "three-eighths of the day", it's there only to confuse you.

10. Hansen and Gretel shared bread crumbs. Hansen took seven-ninths of the crumbs, how much was left for Gretel? *Ans:* two-ninths.

11. A rhesus monkey's brain is one-tenth of the teenager's. By how much the teenager's brain is bigger than the monkey's? *Ans:* by nine-tenths.
Solution: If the teenager's brain is ten-tenths and the monkey's brain is one tenths, then there is a nine-tenths difference between them. I know some teenagers and don't believe there is that much difference.

12. Three blind mice found cheese balls and divided them evenly. What portion of the cheese balls did each mouse get? *Ans:* one-third.
a) What portion did two mice get? *Ans:* two-thirds.

13. Three little pigs and the wolf found a bag of potatoes and shared them (don't be surprised) evenly among themselves. What portion of the bag did three little pigs get? *Ans:* three-fourths, because the whole bag is four-fourths.

14. Mr. Sleeve has a five-pocket jacket. He divided all his money evenly among these pockets. What portion of his money is on two breast pockets? *Ans:* two fifths.

15. Writing the same number of pages every day, Ms. Tale wrote a novel in 7 days. What portion of the pages was left to write after four days? *Ans:* three-sevenths.

16. Dr. Patient split his business cards into 7 equal piles. What portion of the cards is in 2 piles? *Ans:* two-sevenths.
a) If he had 21 cards, how many cards would be two-sevenths? *Ans:* 6 cards.
b) If there were 42 cards, how many cards would be two-sevenths? *Ans:* 12 cards.
c) If he had 49 cards, how many cards would be two-sevenths? *Ans:* 14 cards.

17. Out of 24 socks, two-thirds have holes. How many socks are with holes? *Ans:* 16 socks.

18. If 54 ladybugs sat on 9 grass blades (equal number on each bade), how many ladybugs are on one grass blade?
Ans: 6 lady bugs.
a) How many lady bugs sat on two-ninths of the blades?
Ans: 12 lady bugs.
b) How many lady bugs sat on three-ninths of the blades?
Ans: 18 lady bugs.
c) How many lady bugs sat on five-ninths of the blades?
Ans: 30 lady bugs.

19. If there are 48 raindrops on the window glass, how many drops make three-eighths? *Ans:* 18 drops (48 ÷ 8 = 6, 6 × 3 = 18)

20. A squirrel evenly divided 72 nuts into 8 holes. How many nuts are three-eighths of the holes? *Ans:* 27 nuts.

21. Nazira evenly divided 63 pickles into 9 jars. How many pickles are in five-ninths of the jars? *Ans:* 35 pickles.

22. After picking 54 daisies, Flo divided them into 9 equal bunches. How many daisies are in seven-ninth of the bunches?
Ans: 42 daisies.

23. What is two-fifths plus one-fifth of 25? *Ans:* 15
Solution: Two-fifths plus one-fifth make three-fifths. One-fifth of 25 is 5. Then three-fifths is 15.

24. What is three-seventh plus three-sevenths of 35?
Ans: 30 (or six-sevenths of 30).

25. What is five-eighths plus three-eights of 40?
Ans: 40 (eight-eights or the whole).

26. There were 32 band-aids in a box but after a bike accident only three-eighths of them left. How many are there now?
Ans: 12 band-aids.
Solution: If one-eighth of 32 is 4, then three-eighths is 12.

27. There are 36 utensils in a 6-person silverware set. How many utensils are in a 3-person set? *Ans:* 18 utensils.

28. Mr. Famish ate five-sixths of 48 bagels. How many were left for Mrs. Famish? *Ans:* 8 bagels.
    Solutions: There is an easy way to solve this problem. If Mr. Famish ate five sixths of the bagels, then only one-sixth, or 8 bagels, are left for his wife, Mrs. Famish.

29. Out of 84 tennis balls we only found two-sevenths after practice. How many did we find? *Ans:* 24 balls (one-seventh of 84 is 12, two-sevenths are 24).

30. Out of 54 textbooks in the library, five-ninths were checked out and one-ninth sent out for repair. How many books are out? *Ans:* 36 books or six-ninths of 54 books.

31. Out of 60 feet of wire, an electrician cut three-tenths. How many feet were left after the cut? *Ans:* 42 feet (three-tenths of 60 is 18, them 60 - 18 = 42).

32. If out of $100 birthday money Hugh spent five-tenths, how much is left? *Ans:* $50.

33. Out of 24 awards our team won three-eighths this year. How many awards we didn't win? *Ans:* 15 awards.

34. Gina weighs 54 pounds, Heather is less by one-sixth.
    a) By how many pounds Heather is lighter than Gina?
    Ans: by 9 pounds.
    b) Can you tell Heather's weight? *Ans:* 45 pounds (54 - 9 = 45)

35. There were 50 tea bags in a box. Now, there are two-fifths left. How many bags? *Ans:* 20 bags.

36. Mr. Cal Culator counted 100 peas in a can and took out seven-tenths for his salad. How many peas were left in the can? *Ans:* 70 peas.

37. Mrs. Kelly Culator took three-tenths of 90 kernels on a corn cub for Mr. Culator's salad. How many kernels she left on the cub? *Ans:* 63 kernels.
    Solution: Ten-tenths (the whole) minus three-tenths equals seven-tenths. One tenth of 90 is 9. Then, seven-tenths is 63.

38. Out of 80 beads, Liana used two-tenths for the bracelet and the rest for the necklace. How many beads for the necklace? *Ans:* 64 beads (or eight-tenths).

39. A carpenter loaded 60 nails into a nail gun and used three-tenths for one project and four-tenths for another. How many nails for both? *Ans:* 42 nails (or seven-tenths).

40. Out of 70 sea lions, seven-tenths are in the water. How many are out of the water? *Ans:* 21 sea lions (or three-tenths).

41. Out of 70 sea lions, two-sevenths are babies. How many baby sea lions are there? *Ans:* 20 babies.

42. Out of 90 birds on a plaza two-tenths are pigeons.
    a) How many pigeons are in the plaza? *Ans:* 18 pigeons.
    b) Five-tenths are sparrows. How many? *Ans:* 45 sparrows.
    c) Two-ninths are seagulls. How many? *Ans:* 20 seagulls.
    d) The rest are crows. Caw, caw!

43. Monty cut a loaf of bread into 20 thin slices and ate three-tenths of them. How many did he eat? *Ans:* 6 slices.
    a) How many slices would be six-tenths? *Ans:* 12 slices.

44. Noah also cut his loaf of bread into 20 thin slices and ate three-twentieth of those. How many slices did he eat?
    Ans: 3 slices.
    a) How many slices would be five-twentieths? *Ans:* 5 slices.

45. Olga cut her loaf into 30 thin slices and ate three-thirtieths of them. How many did she eat? *Ans:* 3 slices, of course.

46. In a new 63 light bulb chandelier, two-ninths of the light bulbs burned out. How many light bulbs need to be replaced?
    Ans: 14 light bulbs.

47. There were 80 leaves on a branch. In the fall, eighty-eightieths of the leaves fell down. How many stayed on the branch?
    Ans: Zero leaves.

48. There are 54 aircraft on an Air Force base and six-ninths of them are in the air. How many? *Ans:* 36 aircraft.
    a) How many aircraft are on the ground? *Ans:* 18 aircraft.

49. There are 72 cats and dogs in the rescue shelter and eight-ninths of them are fast asleep. How many are asleep?
    Ans: 64 cats and dogs.
    a) How many are awake? *Ans:* 8 cats and dogs. There is also an elephant, we'll try to ignore the elephant.

50. Two classes, 32 children in each class, were taking a test. Seven-eights of all children finished the test in before the bell. How many did? *Ans:* 56 children.

51. From a 32 oz carton of tomato juice, Mona poured out three-eighths. How many ounces she kept in the carton? *Ans:* 20 ounces.

52. The game lasted 45 minutes, four-fifths of this time Tiger slept on the bench. How much time did stay awake? *Ans:* 9 minutes.

FINDING THE WHOLE NUMBER FROM A FRACTION

I'll tell you how we can find the whole if knowing the fraction of the number. If you know that one-half of a number is 5, then that number is 10 because it takes 2 halves to make the whole, 5 + 5 make 10.

Let's do it again÷ one-third of a number is 1, what is the number? *Ans:* 3, because three-thirds is the whole.

One more time÷ if one-fifth of a number is 3, then the number is 15. That's because when the whole number is divided into five parts and each part is 3, then the whole number is 3 × 5.

EXERCISE I

1. If one-quarter of the number is 3, what's the number?
    Ans: 12, if four-quarters of the number is 3 × 4.

2. If one-quarter of the number is 4, what's the number?
    Ans: 16, if four-quarters of the number is 3 × 4.

3. If one-quarter of the number is 10, what's the number? *Ans:* 40

4. If one-fifth of the number is 5, what's the number? *Ans:* 25

5. If one-half of the number is 20, what's the number? *Ans:* 40

6. If one-seventh of the number is 6, what's the number? *Ans:* 42

7. If one-fifth of the number is 8, what's the number? *Ans:* 40

8. If one-tenth of the number is 7, what's the number? *Ans:* 70

9. If one-fourteenth of the number is 7, what's the number?
   Ans: 98

10. If one-hundredth of the number is 5, what's the number?
    Ans: 500

Now, let's try more fractions:

Problem: Two-fifth of a number is 2, what's the number?
Solution: If two-fifths is 2, then one-fifth is 1.
If one is one=fifth, then the whole number (five-fifths) is 5.

Problem: Two-thirds of a number is 6, what's the number?
Solution: If two-thirds of a number is 6, then one-third is 6 ÷ 2 = 3.
If one-third is 3, then three-thirds (the whole) is 9.

Problem: Two-fifths of a number is 2, what's the number?
Solution: If two-fifths is 2, then one-fifth is 1. If one-fifth is 1,
then the whole number (five-fifths) is 5.

EXERCISE II

1. If two-fifths of a number is 8, what's the number?
   Ans: 20 (one-fifth is 4)

2. If two-thirds of a number is 6, what's the number?
   Ans: 9 (one-third is 3)

3. If three-fifths of a number is 12, what's the number?
   Ans: 20 (one-fifth is 4)

4. If three-quarters a number is 21, what's the number?
   Ans: 28 (one-quarter is 7)

5. If five-eighths of a number is 25, what's the number? *Ans:* 40

6. If three-eighths of a number is 15, what's the number? *Ans:* 40

7. If five-eighths of a number is 30, what's the number? *Ans:* 48

8. If four-ninth of a number is 36, what's the number? *Ans:* 81

9. If three-seventh of a number is 21, what's the number? *Ans:* 49

10. If seven-tenths of a number is 70, what's the number? *Ans:* 100

11. If two-ninths of a number is 18, what's the number? *Ans:* 81

12. If five-ninths of a number is 45, what's the number? *Ans:* 81

13. If four-ninths of a number is 32, what's the number? *Ans:* 72

14. If five-sevenths of a number is 30, what's the number? *Ans:* 42

EXERCISE III

Problem: If three-fourth (or three-quarters) of a number is 27, what's the number?

Solutions: There are two ways to solve this problem.

a) If three-quarters of a number is 27,
the one quarter is $27 \div 3 = 9$.

Then, four-quarters is $9 \times 4 = 36$

b) If three-quarters of a number is 27,
the one quarter is $27 \div 3 = 9$.

Then, three-quarters plus one-quarter is four-quarters, or $27 + 9 = 36$.

Problem: If four-fifths of a number is 32, what's the number?

Solutions: There are two ways to solve this problem.

a) Four-fifths of a number is 32, one-fifth of the number is $32 \div 4 = 8$, and five-fifths (the whole number) is 40.

b) Four-fifths of a number is 32, one-fifth of the number is $32 \div 4 = 8$, and five-fifths of the number (the whole number) is four-fifths plus one-fifths or $32 + 8 = 40$.

Problem: If seven-eighths of a number is 84, what's the number?

Solutions: There are two ways to solve this problem.

a) Seven-eighths of a number is 84, one-eighth is $84 \div 7 = 12$. Then, eight-eighths (the whole number) is $12 \times 8 = 96$.

b) Seven-eighths of a number is 84, one-eighth is
84 ÷ 7 = 12. Then, eight-eighths (the whole number) is
seven-eighths plus one-eighth, or 84 + 12 = 96.

1. If five-sixths of a number is 10, what's the number?
    Ans: 12 (one-sixth is 2)

2. If two-thirds of a number is 24, what's the number? *Ans:* 36
    (one-third is 12)

3. If four-fifths of a number is 16, what's the number? *Ans:* 20

4. If three-fourth of a number is 21, what's the number? *Ans:* 28

5. If four-fifth of a number is 56, what's the number? *Ans:* 70
    (one-fifth is 14)

6. If three-fourth of a number is 33, what's the number? *Ans:* 44

7. If six-seventh of a number is 42, what's the number? *Ans:* 49

8. If seven-eighths of a number is 42, what's the number? *Ans:* 48

9. If eight-ninth of a number is 48, what's the number? *Ans:* 54

10. If nine-tenths of a number is 54, what's the number? *Ans:* 60

11. If ten-eleventh of a number is 100, what's the number?
    Ans: 110

WORD PROBLEMS

1. If 7 students make one-quarter of my class, how big is the
    class? *Ans:* 28 students (if one-quarter is 7 students, then
    four-quarters is 28).

2. Lisa spent $5, or one-quarter of what she earn babysitting, on
    books. How much did she earn? *Ans:* $20 (if one-quarter is $5,
    then four-quarters is $20)

3. In a building only 8, or one-sixth of all windows, are lit. How
    many windows are in the building? *Ans:* 48 windows (if
    one-sixth is 8, then six-sixth is 48).

4. A team sent one-tenth of the athletes to competition. How
    many are on the team, if 9 athletes went? *Ans:* 90 athletes.

5. In school's marching band one-eighth, or 8 students, play drums. How many students are in the band? *Ans:* 64 students.

6. I am thinking of a number. One-ninth of the number is the same as the number of eggs in one dozen. What is the number? *Ans:* 108
   Solution: There are 12 eggs in a dozen. If 12 is one-ninth of the number, then the number is $12 \times 9 = 108$.

7. There are 63 books on a shelf and I have read one-seventh of them. How many did I read? *Ans:* 9 books.

8. A car drove 6 miles, or one-seventh of the distance between two villages. What is the distance between the villages? *Ans:* 42 miles.

9. There are 8 rooms in a house; that's one-ninth of the number of rooms in a palace. How many rooms are in the palace? *Ans:* 72 rooms.
   Palacio Real de Madrid (Royal Palace of Madrid) in Spain is the largest palace in Europe with more than 2,800 rooms. Istana Nurul Iman palace, the official residence of the Sultan of Brunei, is the largest residential palace in the world with 1788 rooms and 257 bathrooms. Forbidden City, Chinese imperial palace in Beijing, is even bigger but it cannot be called a palace as it's made of 980 separate buildings with 8,707 rooms.

10. One-seventh of the stamps in a collection have animals on them. How big is the collection if there are 11 animals' stamps? *Ans:* 77 stamps.

11. There are 28 fish in Sidney's fish tank and one-quarter of them are yellow. How many yellow fish are in the tank? *Ans:* 7 fish.

12. I am thinking of a number. One-seventh of the number is 7. What is the number? *Ans:* 49.

13. I spot some penguins and 6, or one-twelfth of them, are in the water. How many penguins are in and out of the water? *Ans:* 72 penguins, because if 6 penguins are one-twelfth of the total number, then $6 \times 12 = 72$ (penguins).

14. I spot monkeys and 5, or one-fifteenth of them, are babies. How many monkeys do I spot? *Ans:* 75 monkeys (if one-fifteenth is 5, then fifteen-fifteenths is 75).

15. There are 42 tellers in the bank. One-sixth of them work part-time. How many part-time tellers are in the bank? *Ans:* 7 tellers, ask them, they'll tell.

16. Lester bought a box of pens. One-fifth of them, or 10 pens, are red. How many pens are in the box? *Ans:* 50 pens.

17. There were 32 silkworm cocoons and one-eighth of them hatched today. How many hatched? *Ans:* 4 silkworms.

18. A shop ground 8 pounds of coffee, or one-eighth of the stock. What was the stock? *Ans:* 64 pounds.

19. One-tenth of all ships in the navy are cruisers. How many ships are in the navy if there are 11 cruisers? *Ans:* 110 ships.

20. On a balcony in a small Italian village, Signora Vesti hung cloths to dry. One-ninth of all cloths were underpants. How many pieces of cloth did she hang, if there were 7 underpants? *Ans:* 63 pieces of cloth.

21. At a party, one-seventh of all guests went home before dark. How many guests were at the party if 6 guests left early? *Ans:* 42 guests.

22. The Sinker family was fishing. They threw back into the lake 6, or one-twelfth of all fish they caught (too small). How many fish did they catch? *Ans:* 72 fish.

23. The Stinker family took out 6 garbage bags or one-twelfth of all bags. How many bags are there? *Ans:* 72 bags, and that's only in one week.

24. There are 56 miles between the cities and one-eighth of the road is through the tunnel. How long is the tunnel? *Ans:* 7 miles.

25. A farmer picked the grapes and after drying their weight was 12 pounds, or one-quarter of the original weight. How many pounds of grapes did the farmer pick? **Ans:** 48 pounds. Dry grapes are called raisins, you knew that already. What's the name of dry plums? ... Prunes, of course. What about dry tomatoes?... Tomatoes.

26. A railroad inspector checked 8 engines or one-fifteenth of all engines. How many engines were there to be inspected? **Ans:** 120 engines (8 × 15 = 120).

27. If only one-tenth out of 70 darts hit the target, how many missed? **Ans:** 63 throws, because one-tenth (on target) out of 70 throws are 7. Then, 70 - 7 = 63.

28. If one-quarter of 20 plates have chips on them, how many don't? **Ans:** 15 plates.

29. If at the sound of a siren one-ninth of 63 birds flew away, how many stayed? **Ans:** 56 (brave or deaf) birds.

30. If 13, or one-fifth of all potato chips were dipped into a dip, how many went undipped? **Ans:** 52 chips.
    Solution: If 13 chips make one-fifth, then five-fifths (or all chips) are 13 × 5 = 65. Then, 65 (all chips) - 13 (dipped) = 52 (not dipped).

31. We used 4, or one-eighth of all confetti bags for a party. How many bags did we have before the party and how many bags are there now?
    Ans: There were 32 confetti bags before the party and the are 28 bags (32 - 4 = 28) now.

32. A pedestrian walks 3 miles an hour, or one-tenth of the distance a bicyclist rides in the same time. How far can the bicyclist ride in two hours? **Ans:** 60 miles.
    Solution: This is a two part problem. First, we need to find the distance covered by a bicyclist in one hour, which is a 3 miles (pedestrian) × 10 (times more for the bicyclist) = 30 mile in one hour. Then in two hours, the distance is 60 miles.

33. In a 50-hornet nest, one-fifth of all hornets are out. How many are in? *Ans:* 40 hornets.
   Solution: One-fifth of 50 hornets is 10, these are the hornets that are out. That means that 50 - 10 = 40 hornets that are in . Don't go there to check, the hornets don't like visitors.

34. There were 24 chestnuts in the bag and Mr. Castagno roasted one-eighth of them. How many are left in the bag? *Ans:* 21 chestnuts, because one-eighth of 24 is 3, then 24 - 3 = 21.

35. In a caravan of 35 camels and dromedaries, one-fifth are camels. How many dromedaries are in the caravan?
   Ans: 28 dromedaries.
   That's how you tell the difference between them: A dromedary has one hump, long legs and short hair. A camel has short legs, two humps, and long hair.

36. In a pod of 42 dolphins and porpoises, one-sixth is dolphins. How many porpoises are in the pod? *Ans:* 35 porpoises.
   Both, dolphins and porpoises, are mammals, not fish. They have lungs and breathe air. Porpoises are smaller, usually less than 7 feet in length.

37. In a group of 45 rabbits and hares, one-fifth are hares. How many rabbits are in the group? *Ans:* 36 rabbits. Do you want to learn the difference between rabbits and hares? Me too!

38. If one-sixth of 66 fans are Forty-niners, how many root for the other team? *Ans:* 55 fans. San Francisco Forty-niners is a football team.

39. During a short drive Nadine counted 8 red cars, or one-seventh of all cars on the road. How many cars were there? *Ans:* 56 cars.

40. Three-tenth out of 50 CD's in Lim's collection are classical music. How many CD's are classical? *Ans:* 15 CD's.

41. There were 32 fish in a fish tank and Fluffy ate three-eighth of them. How many? *Ans:* 12 fish (if one-eighth of 32 is 4, then three-eighths is 12).

42. In a restaurant, 24 or seven-eighths of all guests used chop-sticks, the rest used forks. How many guests used forks?
   Ans: 3 guests or one-eighth of all guests.

43. A martial art expert broke thee-quarters of 40 wooden boards with his head, and the rest with his hand. How many boards did he break with his hand? *Ans:* 10 boards.

44. One-eighth of all insects in a spider's web are 5 flies. How many insects did it catch? *Ans:* 40 insects.

45. Kirsten paid $8 in taxes for sunglasses, which is one-tenth of the price of the sunglasses. What's the price of the sunglasses? *Ans:* $80. Don't get confused; $8 is one-tenth of the glasses price, not of the whole amount she paid. The whole amount would be $80 (the price of the glasses) + $8 (taxes) = $88.

46. If 12 is one-fifth of a number, what is the number? *Ans:* 60

47. If 13 is one-fifth of a number, what is the number? *Ans:* 65

48. Tamika washed 12 plates or one-fifth of all plates. How many plates were left to wash? *Ans:* 60 plates.

49. Every week 3 students, or one-ninth of the class, help in cafeteria. How big is the class? *Ans:* 27 students.

50. Mr. Bravo donated 6 benches or one-fifth of all benches in the park. How many benches are in the park? *Ans:* 30 benches.

51. One-seventh or 9 frogs in the pond sat on a water lily. How many frogs are in the pond? *Ans:* 63 frogs, not counting the toads. No one wants to count toads?

26

FIND THE NUMBER FROM A FRACTION

If we know that if one-fifth of a number is 4, then the whole number is 20. That is because the whole number is five-fifths, and if one-fifth is 4, it would take five of them to make the whole, or 4 × 5 = 20.

EXERCISE I

1. One-quarter of the number is 12, what's the number ?
   Ans: 48 (four-quarters)

2. One-quarter of the number is 4, what's the number ? *Ans:* 16

3. One-sixth of the number is 18, what's the number ? *Ans:* 108 (six-sixths)

4. One-third of a number is 24, what's the number ? *Ans:* 72

5. One-ninth of a number is 9, what's the number ? *Ans:* 81

6. One-twelfth of a number is 12, what's the number ? *Ans:* 144

7. One-seventh of a number is 14, what's the number ? *Ans:* 98

8. One-fifth of a number is 20, what's the number ? *Ans:* 100

9. One-fifth of a number is 30, what's the number ? *Ans:* 150

10. One-third of a number is 15, what's the number ? *Ans:* 45

11. One-third of a number is 16, what's the number ? *Ans:* 48

12. One-third of a number is 17, what's the number ? *Ans:* 51 (did you notice how the answers increase by 3)

13. One-third of a number is 18, what's the number ? *Ans:* 54

14. One-third of a number is 19, what's the number ? *Ans:* 57

EXERCISE II

If we are told that two-fifths of a number is 8, what's the number? If two-fifths are 8, then one-fifth is half of that or 4. And if one-fifth is 4, then five-fifths is 20.

> *Problem:* One-third of the number is 5, what are two-thirds of the number?
>
> *Solution:* If one-third of the number is 5, the two thirds will be 5 × 2 = 10. *Ans:* 10

> *Problem:* Three-eighths of the number is 12, what's the whole number? What are seven-eighths of the number?
>
> *Solution:* If three-eighths of the number is 12, then one-eighth is 12 ÷ 3 = 4. Then, eight-eighths (the whole number) is 32 and seven-eighths is 28.

1. Two-third of a number is 10. What is the number?
   Ans: 15 (one-third is 5)

2. Two-tenths of a number is 20, what's the number?
   Ans: 100 (one-tenth is 10)

3. Four-tenth of a number is 24, what's the number?
   Ans: 60 (one-tenth is 6)

4. Three-ninths of a number is 9, what's the number?
   Ans: 27 (one-ninth is 3)

5. Two-twelfths of a number is 12, what's the number?
   Ans: 72 (one-twelfth is 6)

6. Five-ninths of a number is 25, what's the number?
   Ans: 45 (one-ninth is 5)

7. One-eighth of a number is 1, what's the number? *Ans:* 8

8. Three-fifths of a number is 3, what's the number? *Ans:* 5

9. Seven-fifteenths of a number is 7, what's the number? *Ans:* 15

10. Nine-ninths of a number is 9, what's the number?
    Ans: 9, of course.

EXERCISE III

Problem: If two-fifths of a number is 8, what's three-fifths?
What's the whole number?

Solution: If two-fifths of a number is 8, then one-fifth is 4.
If one-fifth is 4, then three-fifths is 12. Two-fifths and three-fifths together make five-fifths or the whole. Right?

Then, the whole number is 8 + 12 = 20.

The other way to solve is to find the whole from one-fifth.

Problem: One-seventh of a number is 5, what is two- sevenths?
Solution: Two- seventh s is twice as much as one- seventh, then if one- seventh is 5, then two-seventh s is 10.

Problem: Two-tenths of a number is 6, what is four-tenths?
What's six-tenths of the number?
Solution: If two-tenths are 6, then four-tenths is twice that or 12.
Then six-tenths is three times as much or 18.

1. If one-fifth of a number is 2, what is two-fifths? *Ans:* 4

2. If one-fifth of a number is 2, what is three-fifths? *Ans:* 6

3. If one-fifth of a number is 2, what's four-fifths? *Ans:* 8

4. If three-quarters of a number are 12, what's the number? *Ans:* 16

5. If three-quarters of a number are 24, what's the number? *Ans:* 32

6. If seven-eighths of a number are 35, what's the number? *Ans:* 40

7. If three-eights of a number is 3, what is five-eights? *Ans:* 5

8. If two-fifth of a number is 6, what is four-fifths? *Ans:* 12

9. If three-elevenths of a number is 9, what six-elevenths? *Ans:* 18

10. If two-fifths of a number is 12, what's one-fifth? *Ans:* 6

11. If two-thirteenths of a number is 8, what is one-thirteenth?
    Ans: 4, because one-thirteenth is just a half of two-thirteenths.

12. If four-fifths of a number is 20, what's two-fifths? *Ans:* 10

13. If eight-ninth of a number is 72, what is the number? *Ans:* 81

WORD PROBLEMS

1. A pilgrim divided 14 cabbages into 7 equal piles and gave 3 piles to Roberts family. How many cabbages did Roberts get? *Ans:* 6 cabbages.
   Solution: When all 14 cabbages were divided into 7 piles each pile had 2 cabbages or one-seventh of the whole. When Roberts's received three piles, they have got three-sevenths of the whole, or 6 cabbages.

2. A farmer divided 32 chicken eggs into 8 bowls and sold 3 bowls to a neighbor. How many eggs did he sell? *Ans:* 12 eggs or three-eighths of all eggs.

3. A collector divided 25 coins into 5 holders. How many coins are in 4 holders? *Ans:* 20 coins or four-fifths of all coins.

4. Olivia was told to give two-thirds of her $30 to Petra. How much is that? *Ans:* $20.

5. Petra was told to give three-fifth of her $20 to Reed. How much is that? *Ans:* $12.

6. Reed had to give three-quarters of his $12 to Sid. How much was that? *Ans:* $9.

7. Sid gave two-thirds of $9 to Tom Grundy. How much did Tom get? *Ans:* $6.

8. A class read two-thirds of 21 chapters in the textbook. How many chapters did they read? *Ans:* 14 chapters.

9. A dentist examined three-quarters of 28 teeth. How many teeth did she check? *Ans:* 21 teeth.

10. One-half or 10 students forgot to study for the test. How many students were in the class? *Ans:* 20 student (if one-half is 10, then the whole number is 20).

11. Sofia spent two-thirds of her money, or $12, on the movie tickets. How much money did she have? *Ans:* $18
    Solution: If two-thirds of Sofia's money is $12, then one-third is $6. Then, three-thirds of her money (or all her money) are $6 × 3 = $18.
    You can also say that if two-thirds of Sophia's money is $12 and one-third is $6, then three-thirds, $12 + $6 is $18.

12. Two-thirds of the class, or 14 students, are girls. How big is the class? *Ans:* 21 students.
    Solution: If two-thirds of the class equals 14, then one-third is half of that or 7 students. Then the whole class is 7 (students) × 3 = 21 (students).

13. If three-quarters of all plates, or 9 plates, on the table are empty, how many plates are on the table? *Ans:* 12 plates.
    Solution: If three-quarters are 9 plates, then one-quarter is 3. Then, four-quarters (the whole) is equal to 12 plates.

14. Grandpa Thomas baked 50 pancakes and ate three-tenths of them. How many did he eat? *Ans:* 15 pancakes.
    a) His grand-son Carlos ate four-tenths. How many?
    Ans: 20 pancakes.
    b) His grand-daughter Marisa ate two-tenths. How many?
    Ans: 10 pancakes.
    Marisa also ate the remaining 5 pancakes because she was still hungry.

15. An 18 feet tall tree has branches on two-ninths of its length. How many feet are covered with branches? *Ans:* 4 feet.

16. Veronica wore sandals three-tenths out of the 30 days this month. How many days did she wear sandals? *Ans:* 9 days.
    Solution: The whole number of days is 30, one-tenth of 30 is 3, and three-tenths is 9.

17. A photographer told Serena that only two-fifths of her 25 pictures came out well. How many good pictures were good?
    Ans: 10 pictures.

18. Lillian took three-tenths of the toothpicks from 60-toothpick box for a project. How many toothpicks did she take? *Ans:* 18 toothpicks (one-tenth is 6 and three-tenths is 18).

19. During the field trip Gracie saw 14 sea urchins, but Sissy saw only six-sevenths of that. How many sea urchins did Sissy see? *Ans:* 12 sea urchins.

20. For a bet, a university professor learned 24 languages and promptly forgot five-eighths of them. How many did he forget? *Ans:* 15 languages.
    There are probably more than 6,500 languages spoken in the world. According to Guinness Book of Records, Harold Williams from New Zealand was known to speak over 58 languages fluently as well as some of their dialects.

21. Yarima has 49 pairs of glasses, and three-seventh of them are sunglasses. How many pairs of sunglasses does she have? *Ans:* 21 pairs.

22. In February, Jim went to a gym five-sevenths out of 28 days. How many days did Jim go the gym? *Ans:* 20 days.

23. Seven-ninths of a 45-gallon tank is filled with water. How many gallons of water are in the tank? *Ans:* 35 gallons
    Solution: One-ninth of 45 (gallons) is 5.
    Then seven-ninths is 5 × 7 = 35.

24. A little donkey walked five-eighths of a 64 yard path and stopped. How many yards did it walk? *Ans:* 40 yards.

25. On a full-moon night, a coyote howled 49 times and an owl hooted three-sevenths of that. How many times did the owl hoot? *Ans:* 21 times.

26. A piano player played 54 notes with her right hand and five-sixths of that with her left. How many notes did she play with her left hand? *Ans:* 45 notes.
    a) How many notes did she play with both hands?
    Ans: 99 notes.

27. At noon, the temperature was 63 degrees and now it is eight-ninths of that. What is the temperature now? *Ans:* 56 degrees.

28. There were 50 icicles hanging from the roof; some melted and there are three-tenths of them now. How many? *Ans:* 15 icicles.

29. Out 60 hats in the closet Ms. Bonnet wore only three-tenths last year. How many did she wear? *Ans:* 18 hats.

30. What is seven-tenths of 80? *Ans:* 56.

31. What is seven-tenths of 100? *Ans:* 70.

32. If 100 snowflakes fell down and three-tenths of them melted on my hand, how many didn't? *Ans:* 70 snowflakes (three-tenths of 100 is 30; then, 100 - 30 = 70).

33. There are 30 trees growing on our street. Four-tenths of them are maples. How many maples grow there? *Ans:* 12 maples.
a) Two-tenths of the trees are birches. How many birches? *Ans:* 6 birches.
b) Three-tenths are poplars. How many? *Ans:* 9 poplars.

34. Auntie Chambers put 40 coat hangers in the closet and left three-eighths of them empty for the guests. How many were left for the guests? *Ans:* 15 hangers.

35. Mr. Van den Flor planted 90 tulips, but only nine-tenths of them bloomed. How many bloomed? *Ans:* 81 tulips.
Solution: One way to solve the problem is to find one-tenth of all tulips, 9 tulips and then subtract it from the whole number, because nine-tenths is precisely one-tenth less than the whole.

36. Three-fifths of the children correctly solved all the problems on the test. How many answered correctly if there are 30 children in the class? *Ans:* 18 children.

37. The school library has 40 maps. Four-tenths of them are world maps. How many world maps are there? *Ans:* 16 maps.

38. A store sold 72 fish. Four-ninths of them were frozen. How many frozen fish did they sell? *Ans:* 32 fish.

39. Liz sewed 54 buttons on her cloths but had to remove eight-ninths of them. How many did she have to cut off? *Ans:* 48 buttons. She must have done pretty awful job sewing.
Solution: Eight-ninths is one-ninth less that the whole (nine-ninths). If one-ninth is 54 ÷ 9 = 6, then 54 - 6 = 48 (buttons).

40. Out of 28 parrots in the cage two-sevenths are talking. How many talking parrots are in the cage? *Ans:* 8 talking parrots. There are several parrot species known for their amazing talking abilities. African Greys have big vocabulary and are known to be smart. Quaker Parrots can speak in short phrases and imitate machine sounds. Indian Ringneck Parakeets are attractive and have beautiful voices. In 1995, Puck, a parakeet, was named a bird with the largest vocabulary: 1,728 words.

41. Out of 18 batteries in the pack, Armando used five-sixths for his project. How many batteries did he use? *Ans:* 15 batteries.

42. Ms. Kitchens left 18 sausages on the table and went to answer the phone. Coming back, she found eight-ninths of the sausages missing. How many sausages did her rascally fox terrier Ginger steal from the table? *Ans:* 16 sausages.

43. After eating the sausages, Ginger squealed 21 times. Four-sevenths of them were squeals of delight (the rest were of shame and pain). How many squeals of delight did it make? *Ans:* 12 squeals of delight. Ginger probably got sick after eating that many sausages.

44. If three-quarters or 12 cucumbers on a plate were pickles. How many cucumbers were on the plate? *Ans:* 16 cucumbers. *Solution:* Three-quarters is one-quarter short of the whole (four-quarters). If one-quarter is 12 ÷ 3 = 4 (cucumbers) then four quarters is 12 + 4 = 16.

45. If seven-tenths of all guests at a party, or 21 people, came uninvited, how many people were at the party? *Ans:* 30 people (one-tenth is 3).

46. Halfway to the forest, Hansen dropped three-quarters of 20 bread crumbs. How many were left for the other half of the journey? *Ans:* 5 crumbs. *Solution:* Forget about halfway, that is here only to confuse you. If Hansen dropped three-quarters out of four-quarters of the crumbs, then there is only one-quarter, or 5 crumbs, left.

47. Rapunzel received 42 hair combs for her birthday and broke six-sevenths in one week. How many were left for brushing? *Ans:* 6 combs.

Solution: Take six-sevenths (broken combs) from seven-sevenths (the whole number of combs), that leaves one-seventh of 42 or 6 combs.

48. Two-fifths of all animals in a 30-animal shelter are cats. How many animals are in the shelter? ***Ans:*** 75 animals (because one-fifth is 15).

49. Mrs. Sliva picked 16 plums, or four-fifths of all plums on a branch. How many plums were on the branch? ***Ans:*** 20 plums.

50. Farid counted 18 eggs, or three-tenths of all the eggs, before they hatched. How many eggs were there? ***Ans:*** 60 eggs.

27

SHORT REVIEW OF ADDITION AND SUBTRACTION WITH ELEMENTS OF ALGEBRA

In Verbal Math we add double-digit numbers by adding first tens, next ones, and then adding the sums together. We can also add tens of a number first and then add the remaining ones to the sum.

Problem: 86 + 37 = ?

Solutions: a) 80 + 30 = 110. Next, 6 + 7 = 13. Then, 110 + 13 = 123
 b) 86 + 30 = 116. Then, 116 + 7 = 123
 86 + 37 = 123

105 - 25 = 80	47 + 62 = 109	27 + 79 = 106
49 + 42 = 91	33 + 78 = 111	65 + 58 = 123
64 + 48 = 112	57 + 82 = 139	48 + 44 = 92
64 + 28 = 92	43 + 95 = 138	59 + 79 = 138
73 + 79 = 151	84 + 18 = 102	66 + 19 = 85

In Verbal Math we subtract from double-digit or triple numbers by taking away tens first, and then subtracting ones from the difference. You can also subtract ones first and then tens.

Problem: 115 - 47 = ?
Solution: 115 - 40 = 75. Then, 75 - 7 = 68

Problem: 153 - 83 =?
Solution: 153 - 3 = 150. Then, 150 - 80 = 70.

84 - 28 = 56	103 - 17 = 86	143 - 73 = 70
73 - 57 = 16	123 - 55 = 68	164 - 59 = 105
104 - 44 = 60	114 - 49 = 65	135 - 99 = 36
107 - 85 = 22	153 - 68 = 85	151 - 55 = 96

▶ *A trick: Adding numbers close to 100.* When adding two numbers with one number ending with 9, add 1 to that number and subtract 1 from the other number before adding them together.

Problem: 99 + 154 = ?
Solution: First we add 1 to 99 to make 100 and also taking away 1 from 154 to make it 153. Now, 100 + 153 = 253

You can also try it with numbers ending with 8, but then you will be adding and taking away 2.

Problem: 67 + 98 = ?
Solution: We'll add 2 to 98 to make it 100 and take away 2 from 67 to make it 65. Now, 65 + 100 = 165.

100 + 99 = 199	32 + 98 = 130	98 + 111 = 209
99 + 99 = 198	202 + 99 = 301	102 + 98 = 200
74 + 99 = 173	163 + 98 = 261	66 + 98 = 164
57 + 99 = 156	98 + 99 = 197	109 + 99 = 108

▶ *A trick: Subtracting numbers close to 100.* When subtracting a number ending with 9 from another number, add 1 to both numbers before doing subtraction.

Problem: 121 - 99 = ?
Solution: First, we add 1 to both numbers and get 122 and 100. Then, we do subtraction, 122 - 100 = 22.

You can use this trick subtracting 98 by adding 2 to both numbers.

Problem: 274 - 98 = ?

Solution: First we add 2 to both numbers, they turn into 276 and 100. Then we do subtraction, 276 - 100 = 176.

147 - 99 = 48	198 - 99 = 99	505 - 99 = 406
118 - 99 = 19	401 - 98 = 303	107 - 98 = 9
298 - 99 = 199	135 - 98 = 37	514 - 98 = 416
374 - 98 = 276	366 - 98 = 268	399 - 101 = 298

EXERCISE I

1. The sum of 2 numbers is 104 and one number is 50, what's the other number? *Ans:* 54

2. The sum of 2 numbers is 114 and one number is 99, what's the other number? *Ans:* 15

3. The sum of 2 numbers is 131 and one number is 60, what is the other number? *Ans:* 71

4. The sum of 2 numbers is 144 and one number is 60, what's the other number? *Ans:* 84

5. The sum of 2 numbers is 111 and one number is 98, what's the other number? *Ans:* 13

6. The sum of 2 numbers is 137 and one number is 99, what's the other number? *Ans:* 38

7. The sum of 2 numbers is 108 and one number is 32, what's the other number? *Ans:* 76

8. The sum of 2 numbers is 101 and one number is 49, what's the other number? *Ans:* 52

9. The sum of 2 numbers is 114 and one number is 50, what's the other number? *Ans:* 64

10. The sum of 2 numbers is 145 and one number is 79, what's the other number? *Ans:* 66

11. The sum of 2 numbers is 122 and one number is 61, what's the other number? *Ans:* 61

EXERCISE II

Problem: The sum of 2 numbers is 10 and one number is bigger than the other by 2. What are the numbers?

Solution: We can say that second, larger number is equal to first number plus 2. In other words, 10 (the sum of both numbers) is equal to first number plus another first number plus 2. Then, 10 - 2 is equal to 8, the sum of two equal numbers and half of 8 is 4. Then, one number is 4 and the other number is 6.

Problem: The sum of 2 numbers is 22 and one number is bigger than the other by 4. What are the numbers?

Solution: The second number is equal to first number plus 4. Then, 22 - 4 is equal to 18 or the sum of two first numbers together. Then, 18 ÷ 2 = 9, or one number is 9 and the other number is 9 + 4 = 13.

1. The sum of two numbers is 15, one number is bigger that the other by 3. What are the numbers?
   Ans: 6 and 9 (15 - 3 = 12, 12 ÷ 2 = 6)

2. The sum of two numbers is 31, one number is bigger that the other by 11. What are the numbers?
   Ans: 10 and 21 (31 - 11 = 20, 20 ÷ 2 = 10, 10 + 11 = 21).

3. The sum of two numbers is 43, one number is bigger that the other by 27. What are the numbers?
   Ans: 8 and 35 (43 - 27 = 16, 16 ÷ 2 = 8, 8 + 27 = 35).

WORD PROBLEMS

1. Mr. Anthony had 124 pins in his collection and gave 53 pins to the museum. How many pins are in the collection now?
   Ans: 71 pins.

2. If out of 105 ostrich eggs, 35 hatched. How many didn't?
   Ans: 70 eggs.

3. Out of 105 ostrich eggs, 15 never hatched. How many did?
   Ans: 90 eggs.

4. In a pond, Horace counted 24 frogs and 99 tadpoles. How many creatures did he count? *Ans:* 123 frogs and tadpoles.

5. A chef mixed 64 ounces of milk with 64 ounces of water. How big is the mix? *Ans:* 128 ounces.

6. A chef took 102 ounces of meat and trimmed off 42 ounces of fat. How much meat was left? *Ans:* 60 ounces.

7. A new restaurant ordered 136 forks and used 80 the first day. How many were not used? *Ans:* 56 forks.

8. If there are 133 species of animals on an island and 53 species are birds, how many are non-birds? *Ans:* 80 species.

9. Mr. and Mrs. Lloyd examined their home after an earthquake and found that 45 out their 135 vases were broken. How many survived the earthquake? *Ans:* 90 vases.

10. Mr. Smirnoff was shocked when he looked at the mushrooms his son picked in the woods, 80 out of 104 were poisonous. How many good mushrooms were there? *Ans:* 24 mushrooms.

11. A military base has 77 helicopters and 77 planes. How many aircraft does the base have? *Ans:* 154 aircraft.

12. A musical store sold 56 violin strings and 56 cello strings. How many strings were sold? *Ans:* 112 strings.

13. An old lady had 112 cats at her house, 62 of them were males. How many female cats did she have?
Ans: 50 female cats. Canadians Jack & Donna Wright made their way to the Guinness Book of Records for having 689 cats.

14. Before his voyage, Sinbad the Sailor weighed 157 pounds. While traveling he gained 36 pounds. How much did he weigh after the trip? *Ans:* 193 pounds.

15. During his voyage, Sinbad the Sailor tried to catch 104 exotic animals but caught only 28. How many got away?
Ans: 76 animals.

16. During his voyage, Sinbad the Sailor tried to catch 104 exotic birds but caught only 77. How many got away? And: 27 birds. He was luckier with birds.

17. During his voyage, Sinbad the Sailor met 141 strangers and 99 of them tried to quarrel with Sinbad. How many didn't?
Ans: 42 (friendly) strangers.

18. One office building has 135 offices but only 98 are occupied. How many are empty? *Ans:* 37 offices.

19. One composer wrote 68 songs, and another wrote 37 more songs. How many songs did both write?
Ans: 173 songs (68 + 37 = 105, then 68 + 105 = 173)

20. Out of 173 songs both composers wrote, 98 songs are in English and the rest are in Spanish. How many Spanish songs did they write? *Ans:* 75 songs.

21. There are 75 species of mushrooms in a forest and 46 species of fern. How many species of both grow in the forest?
Ans: 121 species.

22. The young White Rhinoceros' horn is 54 centimeters. That is 57 centimeters shorter than the old rhino's horn. How long is the old rhino's horn? *Ans:* 111 centimeters.
White Rhinoceroses live in Africa. The record-sized White Rhinoceros was about 4,600 kg (10,000 lb). They are now considered extinct because none have been seen for many years.

23. Unrefrigerated cranberry juice can stay fresh for 19 days and refrigerated juice can be used after 104 days. How much longer can the refrigerated juice stay? *Ans:* 85 days.

24. It takes 45 ounces of metal to make a small bell and 87 ounces more for a big bell. How much metal is needed for both bells?
Ans: 177 ounces (45 + 87 = 132, then 45 + 132 = 177).

25. One knight's sword is 78 inches, another knight has 83 inch sword. By how much the second sword is longer than the first?
Ans: 5 inches.

26. One composer wrote an opera in 122 days. Another, in 56 days. How much longer did it take the first composer to write an opera? *Ans:* 66 days more.

27. It takes 112 millimeters of wire to make a large paperclip and 47 millimeters less for a small one. How much wire does it take for the small paperclip? *Ans:* 65 millimeters.

28. After a hailstorm, Denise counted 57 dents on mom's car and 67 dents on brother's truck. How many dents did she count?
Ans: 124 dents. Hailstones can be the size of a softball.

29. A small hailstone may fall at a speed of 43 miles per hour, a large one at 111 miles per hour. How much faster is the larger stone? *Ans:* 68 mph.

30. At 7 AM, a shadow from a pole is 122 inches and at 10 AM it's 66 inches shorter. What is the length of the shadow at 10 AM? *Ans:* 56 inches.

31. When 105 students went on a field trip to an apiary (bee-garden), only 66 students tried honey. How many didn't? *Ans:* 39 students.

32. A basket of fruit weighed 176 ounces. After the fruits were dried, it weighed 99 ounces. How much weight was lost? *Ans:* 77 ounces.

33. A herpetologist (a scientist who studies snakes) described 115 species of poisonous snakes in Central America and 89 species in North America. How many more Central American species did she describe? And: 26 species.

34. A family that was using 153 gallons of water per day decided to reduce what they use by 65 gallons. How much water do they use now? *Ans:* 88 gallons. A family of four might use more than 1,000 gallons of water a day if they are not trying to save it.

35. The length of Pandora's box is 103 inches. It's 47 inches longer than its height. What's its height? *Ans:* 56 inches.

36. Arno paid $1 and 12¢ (112¢) for two items. One item costs 89¢. What's the price of the other item? *Ans:* 23¢.

37. A snowboard and boots together cost $177, and the boots alone cost $78. What's the snowboard's price? *Ans:* $99.

38. An airline ticket costs $160 and one day in a hotel is $80. How much do both cost? *Ans:* $240.

39. The tickets and the ride to the circus cost $171. The ride alone cost $89. How much are the tickets? *Ans:* $82.

40. The price of two tickets is $82. The adult ticket is $56. How much is the child's ticket? *Ans:* $26.

41. A small can of night crawlers costs 79¢ and the large can of worms is 140¢ more. How much does the larger can cost? **Ans:** 219¢ or 2 dollars and 19¢.

42. One puppeteer has 55 puppets and the other has 45. How many do both have? **Ans:** 100 puppets.

43. In at the market in New Delhi a watermelon costs 65 rupees. How much do 2 watermelons cost? **Ans:** 130 rupees.

44. At the same market, a chicken costs 87 rupees. What's the price of 2 chickens? **Ans:** 174 rupees.

45. Two retired generals together have 55 decorations. One general has 15 more than another. How many decorations does each have? **Ans:** 20 and 35 decorations.

46. An office received 105 pieces of mail, letters and parcels. There were 3 more letters than parcels. How many parcels did they receive? **Ans:** 51 parcels.

47. Out 156 grapes, 59 were sour. How many were not sour grapes? **Ans:** 97 grapes.

48. Rusty took $148 to the sheep auction and bought a $75 sheep. How much money was left to buy another sheep? **Ans:** $73.

49. In the past, grain was measured by bushels. If a farmer collected 166 bushels of corn and sold 98 bushels, how many bushels were left? **Ans:** 68 bushels.
Bushel is a measure of volume and equal to approximately 8 gallons of dry goods.

50. A shop has 173 pounds of mocha and java coffee. How many pounds of java are there if there are 53 pounds more mocha than java? **Ans:** 60 pounds of java.

28

DIVISION OF DOUBLE AND TRIPLE-DIGIT NUMBERS

Division of double and triple digit numbers might look difficult, but, in fact, it is fairly easy if you remember addition and subtraction facts and also multiplication table.

Problem: $68 \div 4 = ?$

▶ *A trick:* These problems are easy if we separate the number into two or more parts and divide each part separately. It's always easy to divide a number by its multiple of ten.

$$50 \div 5 = 10$$
$$70 \div 7 = 10$$
$$90 \div 9 = 10$$
$$100 \div 10 = 10$$
$$130 \div 13 = 10$$

That's how we do it:

First, we took a divider (4) and multiplied it by 10. Then, we subtracted the product (40) from the number we divide (68) to get 28. By now, you should be able to do all this in your head.

Solution: We split 68 into 40 and 28 and then divide each number separately by 4. Then, we only need to add the results together. In our problem $40 \div 4 = 10$ and $28 \div 4 = 7$. Next, $10 + 7 = 17$.

Problem: $88 \div 4 = ?$

Solution: 88 can be split into $40 + 40 + 8$. Next, $40 \div 4 = 10$; another $40 \div 4 = 10$; and $8 \div 4 = 2$. Then, $10 + 10 + 2 = 22$

$88 \div 4 = 22$

Problem: $117 \div 9 = ?$

Solution: The number 117 is equal to $90 + 27$. Next, $90 \div 9 = 10$ and $27 \div 9 = 3$. Then, $10 + 3 = 13$.

Let's do one more problem together.

Problem: $108 \div 6 = ?$

Solution: $108 = 60 + 48$. Next, $60 \div 6 = 10$ and $48 \div 6 = 8$. Finally, $10 + 8 = 18$

$108 \div 6 = 18$

EXERCISE I

$24 \div 8 = 3$	$48 \div 2 = 24$	$91 \div 7 = 13$
$24 \div 4 = 6$	$88 \div 2 = 44$	$84 \div 6 = 14$
$52 \div 3 = 14$	$48 \div 3 = 16$	$44 \div 2 = 22$
$48 \div 8 = 6$	$51 \div 3 = 17$	$44 \div 4 = 11$
$36 \div 3 = 12$	$57 \div 3 = 19$	$66 \div 6 = 11$
$66 \div 3 = 22$	$52 \div 4 = 13$	$48 \div 3 = 16$
$69 \div 3 = 23$	$75 \div 5 = 15$	$55 \div 5 = 11$
$82 \div 2 = 41$	$78 \div 6 = 13$	$72 \div 6 = 12$

EXERCISE II

48 ÷ 2 = 24	92 ÷ 4 = 23	91 ÷ 7 = 13
52 ÷ 2 = 26	84 ÷ 4 = 21	85 ÷ 5 = 17
50 ÷ 2 = 25	84 ÷ 6 = 14	78 ÷ 3 = 23
66 ÷ 3 = 11	84 ÷ 7 = 12	76 ÷ 4 = 19
45 ÷ 3 = 15	93 ÷ 3 = 31	98 ÷ 7 = 14
69 ÷ 3 = 13	90 ÷ 6 = 15	84 ÷ 4 = 21
65 ÷ 5 = 13	92 ÷ 4 = 23	96 ÷ 4 = 24
88 ÷ 4 = 22	96 ÷ 8 = 12	81 ÷ 3 = 27

EXERCISE III

Dividing triple-digit numbers by one-digit number is very much like dividing double-digit numbers.

- First, look at the divider and mentally multiply it by 10.
- Next, subtract the product (10 × the divider) from the number you divide.
- Then, divide two numbers separately by the divider (of course you know the first answer already, it's 10).
- Finally, add 10 and the result of the other number division together).

Look:

Problem: 102 ÷ 6 = ?
Solution: First, 6 × 10 is 60. Next 102 is equal to 60 + 42. Then, 60 ÷ 6 = 10 and 42 ÷ 6 = 7. Together, it's 17.

102 ÷ 6 = 17

Problem: 105 ÷ 7 = ?
Solution: 105 is equal to 70 + 35. Then, 70 ÷ 7 = 10 and 35 ÷ 7 = 5.

10 + 5 = 15.

120 ÷ 6 = 20 (120 is equal to 60 + 60; then 60 ÷ 6 = 10 and 10 + 10 = 20)

Did you guess that 180 ÷ 6 = 30?

Because 180 is equal to 60 + 60 + 60.

111 ÷ 3 = 37 (111 is equal to 90 + 21.
Then, 90 ÷ 3 = 30 and 21 ÷ 3 = 7.

Together, 30 + 7 = 37)

112 ÷ 4 = 28 (112 is equal to 80 + 32. Next, 80 ÷ 4 = 20 and
32 ÷ 4 = 8. Then, 20 + 8 = 28)

110 ÷ 10 = 11	112 ÷ 8 = 14	120 ÷ 10 = 12
102 ÷ 3 = 31	112 ÷ 7 = 16	171 ÷ 9 = 19
104 ÷ 8 = 13	128 ÷ 8 = 16	150 ÷ 10 = 15
105 ÷ 5 = 21	126 ÷ 7 = 18	200 ÷ 10 = 20
105 ÷ 7 = 15	136 ÷ 8 = 17	132 ÷ 6 = 22
126 ÷ 9 = 14	126 ÷ 9 = 14	133 ÷ 7 = 19
108 ÷ 4 = 26	133 ÷ 7 = 19	140 ÷ 7 = 20
120 ÷ 8 = 15	144 ÷ 9 = 16	
99 ÷ 9 = 11	152 ÷ 8 = 19	
108 ÷ 9 = 12	135 ÷ 9 = 15	

WORD PROBLEMS

1. A squirrel evenly divided 44 nuts among 4 tree hollows. How many nuts are in each hollow? *Ans:* 11 nuts.

2. Another squirrel divided 33 acorns among 3 tree hollows. How many in each hollow? *Ans:* 11 acorns.

3. A base guitar has 4 strings. How many guitars can you string with 52 strings? *Ans:* 13 base guitars.

4. If 52 buttons were sewn on 4 shirts, how many buttons were on each shirt? *Ans:* 13 buttons.

5. If total numbers of legs were counted at 88, how many octopi were counted? *Ans:* 11 octopi.
   I've heard a story about a pet store owner who had to order 6 octopi. He wasn't sure whether to write octopuses or octopi. To avoid embarrassment he wrote: "Please send me one octopus, and while at it, please send 5 more."

6. How many tricycles would have 42 wheels? *Ans:* 14 tricycles.

7. How many $5 bills make $60? *Ans:* 12 bills.

8. If triplets together weigh 39 pounds, how much does each weigh? *Ans:* 13 pounds.

9. How many 3-headed dragons did the knights fight, if together they had 51 heads (the dragons, of course, not the knights)? *Ans:* 17 dragons, an odd number if you ask me.

10. If a family of 7 paid $119 for concert tickets, what was the price of each ticket? *Ans:* $17

11. If 70 mangoes were divided among 5 boxes, how many are in each box? *Ans:* 14 mangoes.

12. If 96 strings were strung on six string guitars, how many guitars were strung? *Ans:* 16 guitars.

13. A Russian balalaika has 3 strings. How many balalaikas could you sting with 57 strings? *Ans:* 19 balalaikas.

14. Nora used 52 slices of cheese to make 4 sandwiches. How many slices went to each sandwich? *Ans:* 13 slices.

15. A coyote has 4 legs. How many coyotes have 60 legs? *Ans:* 15 coyotes.

16. A violin also has 4 strings. How many violins can be strung with 56 strings? *Ans:* 14 violins.

17. Russian guitar has 7 strings. How many Russian guitars can one string with 84 strings? *Ans:* 12 guitars.

18. How many squares have 44 corners? *Ans:* 11 squares, because each has 4.

19. Each volleyball team has 6 players. How many teams do 48 players make? *Ans:* 8 teams.

20. If a team has 6 players, how many teams does 96 players make? *Ans:* 16 teams.

21. How many weeks are in 84 days? *Ans:* 12 weeks.

22. How many weeks are in 105 days? *Ans:* 15 weeks.

23. How many weeks are in 140 days? *Ans:* 20 weeks, because 140 (weeks) is equal to 70 + 70.

24. How much does each egg weigh if 8 eggs weigh 120 ounces? *Ans:* 15 ounces.

25. How much does each egg cost if 7 eggs cost 98¢? *Ans:* 14¢.

26. How many feathers are in each parrot's tail if 8 parrots have 136 tail feathers? *Ans:* 17 feathers (136 is equal to 80 + 56).

27. If one book has 24 pages, how many pages are in 3 books? *Ans:* 72 pages.

28. How many miles did a horse run each hour, if in 6 hours it covered 90 miles? *Ans:* 15 miles.

29. If a shop repaired 9 grandfather clocks in 108 days, how many days, on average, did it take to fix each clock? *Ans:* 12 days.

30. If a caterpillar ate 119 leaves in 7 hours, how many leaves did it eat each hour? *Ans:* 17 leaves (70 + 49).

31. If a silkworm made 133 millimeters of silk in 7 hours, how long is a silk line made in one hour? *Ans:* 19 millimeters (70 + 63).

32. If a diver took 144 breaths breathing 8 breaths per minute, how many minutes did he breathe? *Ans:* 18 minutes (80 + 64).

33. If I cut a 102 feet of rope into 6 equal pieces, what's the length of each piece? *Ans:* 17 inches (60 + 42).

34. If 152 chocolates were divided evenly among 8 boxes, how many were in each box? *Ans:* 19 chocolates (80 + 72).

35. If a printer prints 5 pages per minute, how many pages will it print in 15 minutes? *Ans:* 75 pages.

36. A kangaroo covered 126 feet in 9 jumps. How long was each jump? *Ans:* 14 feet.

37. If 8 rattle snakes together laid 104 eggs, how many did each snake lay, if each snake laid the same number of eggs? *Ans:* 13 eggs.

38. If one monkey eats 4 pounds of bananas a day, how many monkeys can you feed with 76 pounds of bananas? *Ans:* 19 monkeys.

39. If a donkey eats 8 pounds of grass a day, how many days can you feed it with 120 pounds of grass? *Ans:* 15 days.

40. If one zebra eats 9 pounds of grass, how many zebras can you feed with 135 pounds of grass? *Ans:* 15 zebras.
    Did you know that zebras and donkeys can mate and their offspring is called a zonkey, zedonk, or zebrass?

41. If one peacock weighs 9 pounds, how many peacocks weigh 126 pounds? *Ans:* 14 peacocks.

42. If one household uses 6 gallons of milk a week, how many households use 90 gallons a week? *Ans:* 15 households.

43. If a quarter coin weighs 6 grams, how many coins weigh 108 grams? *Ans:* 18 coins.

44. If a dollar coin weighs 8 grams, how many coins weigh 136 grams? *Ans:* 17 coins.

45. A farmer collected 72 bushels (remember bushels?) of wheat and divided them into 4 bins. How many bushels are in each bin? *Ans:* 18 bushels.

46. If it takes Alan 13 minutes to take his shower and brush the teeth every day, how much time does it take him in one week (7 days)? *Ans:* 91 minutes (more than an hour and a half).

47. If a faucet runs water at the rate of 14 gallons per minute, how many gallons will it pour out in 8 minutes? *Ans:* 112 gallons.

48. Another faucet's flow is 112 gallons in 7 minutes. How many gallons of water flow each minute? *Ans:* 16 gallons.

49. A lady gave a baker a $120 and asked for all the $8 cakes her money can buy. How many cakes did she buy? *Ans:* 15 cakes.

FACTORS AND ELEMENTS OF ALGEBRA

REVIEW OF FACTORS

Numbers you multiply together are called *factors*. Factor in arithmetic is any number that divides a given number evenly. Factors are two or more numbers when multiplied present the given number. A number can be a product of two or more factors. Finding all factors for a number is easier than it looks if you remember your times table.

Use your Verbal Math tricks:

- Start with 1, it's always a factor.
- Then, 2. If your number is even, then 2 is a factor. If not, it isn't.
- If sum of digits in your number is divided by 3, then the number is divided by 3.

Question: Is 3 a factor for 123?
Ans: Yes, because 1 + 2 + 3 = 6 and 6 is divided by 3.

Question: Is 3 a factor for 221?
Ans: No, 2 + 2 + 1 = 5 and 5 is not divided by 3.

If the number can be divided by 2 once and again, then 4 is a factor.

Only numbers that end with 0 and 5 can be divided by 5.

If sum of digits in your number is divided by 9, then the number is divided by 9.

Let's try. What are the factors of 28?
Solution: 1 is always a factor. Also 2 (because the number is even) and 14 (because $28 \div 2 = 14$). Not 3, because $2 + 8 = 11$, but 4 is the factor (because $28 \div 4 = 7$), and also 7. And, of course, 28.
Ans: Factors of 28 are 1, 2, 4, 7, 14, and 28.

Each number has at least 2 factors: 1 and itself. For number one both factors are the same, 1.

Number 2 also has 2 factors: 1 and 2, as number 3 (1 and 3).

As you might have guessed, all prime numbers have 2 factors, because by definition, prime is the number that only divides by 1 and itself.

Number 4 has three factors: 1, 2, and 4.
Number 6 has four factors: 1, 2, 3, and 6.

How many factors does number 12 have? **Ans:** six factors: 1, 2, 3, 4, 6, 12 because you can divide number 12 by any of these numbers without remainder.

1. What are the factors of 4? **Ans:** 1, 2, 4

2. What are the factors of 8? **Ans:** 1, 2, 4, 8

3. What are the factors of 11? **Ans:** 1, 11

4. What are the factors of 14? **Ans:** 1, 2, 7, 14

5. What are the factors of 16? **Ans:** 1, 2, 4, 8, 16

6. What are the factors of 17? **Ans:** 1 and 17

7. What are the factors of 22? **Ans:** 1, 2, 11, 22

8. What are the factors of 26? **Ans:** 1, 2, 13, 26

9. What are the factors of 30? **Ans:** 1, 2, 5, 10, 15, 30

10. If we can divide 12 by 4 without a remainder, then 4 is factor of 12.

11. If we can multiply 3 by 4 to get 12, then both, 3 and 4 are factors.

EXERCISE I

1. How many factors does number 9 have? **Ans:** three, 1, 3, and 9.

2. How many factors does number 11 have? **Ans:** two, 1 and 11. It's a prime number.

3. How many factors does number 14 have? **Ans:** four, 1, 2, 7, and 14.

4. How many factors does number 15 have? **Ans:** four, 1, 3, 5, and 15.

5. How many factors does number 18 have? **Ans:** five, 1, 2, 3, 6, 9, and 18.

6. How many factors does number 20 have? **Ans:** six, 1, 2, 4, 5, 10, and 20.

7. How many factors does number 23 have? **Ans:** two, 1 and 23. It's a prime number.

8. How many factors does 24 have? **Ans:** eight, 1, 2, 3, 4, 6, 8, 12, and 24.

9. How many factors does number 25 have? **Ans:** three, 1, 5, and 25.

10. How many factors does number 28 have? **Ans:** six, 1, 2, 4, 7, 15, and 28.

EXERCISE II

A number can have several pairs of factors (multipliers). For example, number 12 has at least three pairs÷ 1 and 12; 2 and 6; 3 and 4.

1. The product of two numbers is 16. What are the factors combinations? **Ans:** 1 and 16; 2 and 8; and 4 and 4.

2. The product of two numbers is 18, what are the pairs of multipliers? **Ans:** 1 and 18; 2 and 9; 3 and 6.

3. The product of two numbers is 24, what are the pairs of multipliers? *Ans:* 1 and 24; 2 and 12; 3 and 8; 4 and 6.

4. The product of two numbers is 20, what are the pairs of multipliers? *Ans:* 1 and 20; 2 and 10; 4 and 5.

5. The product of two numbers is 26, what are the pairs of multipliers? *Ans:* 1 and 26; 2 and 13.

6. The product of two numbers is 33, what are the pairs of multipliers? *Ans:* 1 and 33; 3 and 11.

7. The product of two numbers is 30, what are the pairs of multipliers? *Ans:* 1 and 30; 2 and 15; 3 and 10; 5 and 6.

8. The product of two numbers is 42, what are the pairs of multipliers? *Ans:* 1 and 42; 2 and 21; 3 and 14; 4 and 13; 6 and 7.

9. The product of two numbers is 46, what are the pairs of multipliers? *Ans:* 1 and 46; 2 and 23.

10. The product of two numbers is 54, what are the pairs of multipliers? *Ans:* 1 and 54; 2 and 27; 3 and 18; 6 and 9.

11. The product of two numbers is 50, what are the pairs of multipliers? *Ans:* 1 and 50; 2 and 25; 5 and 10.

EXERCISE III

Common factors are the numbers by which two or more other numbers can be divided evenly, without remainder.

For numbers 6 and 9, for example, 3 is the common factor because both can be divided by 3.

Common factor for 18 and 21 is also 3. The numbers 12 and 18 have several common factors: 2, 3, and 6. The numbers 6, 15, and 21 have only one common factor - 3.

Note: number 1 should not be considered a common factor.

1. What are the common factors for 42 and 49? *Ans:* 7

2. What are the common factors for 28 and 4? *Ans:* 4

3. What are the common factors for 32 and 28? *Ans:* 2 and 4

4. What are the common factors for 31 and 35? *Ans:* none, and 31 is a prime number.

5. What are the common factors for 22 and 44? *Ans:* 2 and 11.

6. What are the common factors for 45 and 36? *Ans:* 9

7. What are the common factors for 18 and 27? *Ans:* 3 and 9

8. What are the common factors for 25 and 55? *Ans:* 5

9. What are the common factors for 14 and 56? *Ans:* 2 and 7

10. What are the common factor of 23 and 39? *Ans:* 13

EXERCISE IV

We can find the GCF, *greatest common factor* (also known as *greatest common divisor*), of two or more numbers.

For example, numbers 16 and 24 have several common factors: 2, 4, and 8. Their greatest common factor, however, is 8. That means that there is no number larger than 8 by which both numbers can be divided.

The numbers 30, 50, and 60 have the greatest common factor 10.

1. What is the greatest common factor of 25, 50 and 75? *Ans:* 25

2. What is the greatest common factor of 30, 45? *Ans:* 15

3. What is the greatest common factor of 22 and 33?
   Ans: 11, the greatest and the only.

4. What is the greatest common factor of 24 and 40? *Ans:* 8

5. What is the greatest common factor of 18, 27, and 63? *Ans:* 9

6. What is the greatest common factor of 28 and 42? *Ans:* 14

7. What is the greatest common factor of 17, 51 and 85? *Ans:* 17

8. What is the greatest common factor of 56, 64 and 81?
   Ans: none

9. What is the greatest common factor of 32 and 88? *Ans:* 8

10. What is the greatest common factor of 40, 72 and 96? *Ans:* 8

▶ *A trick:* *Multiplying double-digit numbers by 9.* When multiplying double-digit number by nine, multiply the number by 10 and then subtract the number from the product.

Problem: $15 \times 9 = ?$
Solution: First, multiply 15 by 10, $15 \times 10 = 150$.
Then, $150 - 15 = 135$.

$15 \times 9 = 135$

Problem: $34 \times 9 = ?$
Solution: $34 \times 10 = 340$, then $240 - 34 = 206$

$34 \times 9 = 206$

$14 \times 9 = 126$	$25 \times 9 = 225$	$34 \times 9 = 306$
$18 \times 9 = 162$	$28 \times 9 = 252$	$38 \times 9 = 342$
$23 \times 9 = 207$	$31 \times 9 = 279$	$27 \times 9 = 243$
$17 \times 9 = 153$	$33 \times 9 = 287$	$39 \times 9 = 351$

A very similar trick works for double-digit numbers that end with 9 multiplied by a single-digit number. Here we convert the double-digit number to the nearest multiple of ten (the number that ends with 0) and after multiplying it subtract the single-digit number from the product.

Problem: $19 \times 4 = ?$
Solution: First, convert 19 to 20. Next, multiply 20 by 4, $20 \times 4 = 80$. Then, subtract 4 from the product, $80 - 4 = 76$.

$19 \times 4 = 76$

Problem: $39 \times 7 = ?$
Solution: Convert 39 to 40. Now, $40 \times 7 = 280$.
Next, $280 - 7 = 273$.

$39 \times 7 = 273$	$29 \times 6 = 174$	$49 \times 6 = 294$
$19 \times 7 = 133$	$29 \times 8 = 232$	$59 \times 3 = 177$
$19 \times 5 = 95$	$39 \times 5 = 195$	$59 \times 5 = 295$
$19 \times 9 = 171$	$39 \times 8 = 312$	
$29 \times 4 = 116$	$39 \times 9 = 351$	

WORD PROBLEMS

1. How many hungry hunters can evenly divide 12 slices pizza?
   Ans: It can be either 2, or 3, or 4, or 6, or 12. It can also be 1 hunter too, I guess.

2. Among how many pockets one can evenly divide 18 coins?
   Ans: 1, 2, 3, 6, 9, and 18 pockets.

3. What's the smallest number, other than 1, by which we can divide 52 and 72? **Ans:** 3.

4. Remember the rule: If the sum all digits in a number can be divided by 3, then the whole number is divisible by 3.

5. Which one of these numbers can be divided by 3: 27; 112; 123; 462; 612; 729; 609; 1,053? **Ans:** All of them, except for 112.

6. How many students can evenly divide 56 French fries?
   Ans: either 2, or 4, or, 7, or 8, or 14, or 28, or 56 students. In the last case, there will be 56 very hungry students.

7. Among how many fingers can a lady evenly divide her 28 rings?
   Ans: either 2, or, 4, or 7 fingers. Personally, I don't know any lady who has 14 or 28 fingers.

8. How many tourists can 51 hungry mosquitoes sting, if the mosquitoes always divide evenly for their meals? **Ans:** either 3, or 17, or 51 tourists.

9. By what numbers can we divide both, 12 and 14? **Ans:** 2.

10. By what numbers can we divide both, 16 and 26? **Ans:** 2.

11. By what numbers can we divide both, 24 and 36?
    Ans: 2, 3, 4, 6, 12.

12. By what numbers can we divide both, 15 and 45?
    Ans: 3, 5, and 15.

13. By what numbers can we divide both, 32 and 24? **Ans:** 2, 4, 8.

14. By what numbers can we divide both, 42 and 63?
    Ans: 3, 7, and 21.

15. How many bunches can you make with 32 flowers?
    Ans: two 16 flower bunches, four 8 flower bunches, eight 4 flower bunches, and sixteen 2 flower bunches. We can't make a bunch with only one flower.

16. How many equal teams can 26 firefighters make?
    Ans: either two 13 person teams or thirteen 2 people teams. We can't make a team with only one firefighter.

17. How many matching shirts can one make with 20 buttons?
    Ans: two 10 button shirts, four 5 button shirts, five 4 button shirts, and ten 2 button shirts. I guess you also make a one-button shirt.

18. By what numbers can we divide both 42 and 63? *Ans:* 3, 7, and 21.

19. Name all double-digit numbers divisible by 8?
    Ans: 16, 24, 32, 40, 48, 56, 64, 72, 80, 88, 96.

20. Name all the double-digit numbers divisible by 9?
    Ans: 18, 27, 36, 45, 54, 63, 72, 81, 90.

21. How else can one arrange squares on 64 squares (8×8) chess board (except, of course, having all 64 squares in one row)?
    Ans: 32×2 and 16×4.

22. If one asked to break a dollar bill into even parts, what are possible combinations other than all pennies?
    Remember your coins.
    Ans: 2 half dollars (50¢), or 4 quarters, 10 dimes, or 20 nickels.

23. How many ways can a colonel arrange 30 medals on his chest (except of course to put them all in one row)?
    Ans: 2 rows with 15 medals in each row; 3 rows of 10; 5 rows of 6; 6 rows of 5; 10 rows of 3; and 15 rows of 2. Colonels must have big chests.

24. The numbers 26 and 24 have 2 as a common factor? What are other common factors for these numbers? *Ans:* none, only 2.

25. The numbers 27 and 36 have a common factor 9? What are other common factors for these numbers? *Ans:* 3.

26. The numbers 16 and 24 have a common factor 8? What are other common factors for these numbers? *Ans:* 2 and 4.

27. The numbers 33 and 66 have common factors. What are they? *Ans:* 3, 11 and 33.

28. The numbers 77 and 63 have common factors. What are they? *Ans:* only 7.

29. Among how many jars can a chemist divide 32 oz of a liquid? *Ans:* either 2, or 4, or 8, or 16, or 32 jars.

30. What's the *greatest common factor* for both, 24 and 32? *Ans:* 8.

31. What's the *greatest common factor* for both, 30 and 42? *Ans:* 6

32. What's the *greatest common factor* for both, 28 and 35? *Ans:* 7

33. What's the *greatest common factor* for both, 28 and 42? *Ans:* 14

34. What's the *greatest common factor* for both, 36 and 45? *Ans:* 9

35. What's the *greatest common factor* for both, 36 and 54? *Ans:* 18

36. A store received 9 boxes with 20 new cell phones in each box. One phone in each box was broken. How many good phones were there? *Ans:* 171 phones.
    Solution: There are two ways to solve this problem.
    a) First, we find the number of all phones received by the store, $20 \times 9 = 180$ (phones). Then, we find bad phones, 1 (bad phone) $\times$ 9 (boxes) = 9 (bad phones). Finally, we subtract bad phones from all phones, 180 - 9 = 171 (good phones).
    b) First, we "remove" bad phones from each box, 20 - 1 = 19 (good phones in each box). Then, we find all good phones, 19 (phones) $\times$ 9 (boxes) = 171 phones.

37. In class, 19 students received 6 summer assignments each. How many assignments did they all get? *Ans:* 114 assignments.

38. A building has 5 flights of stairs, with 39 steps in each flight. How many steps are in all? *Ans:* 195 steps. Why calling them flights if they don't move and others have to walk them?

39. If a car tank can be filled with 9 gallons of gas, how many gallons will fill 34 tanks? *Ans:* 306 gallons.

40. A math quiz has 29 problems. How many problems did the teacher look at after checking one-third of 24 tests?
    Ans: 232 quizzes.
    Solution: this is a two-part problem. First, we need to find how many quizzes the teacher checked, one-third of 24 is 8. Then, we need to find the total number of problems, 29 (problems in each quiz) × 8 = 232.

41. If a squirrel can crack 9 nuts a minute, how many nuts can it crack in 25 minutes? *Ans:* 225 nuts.

42. If a shoe shiner shines 28 shoes a day, how many shoes can he shine in 9 days? *Ans:* 252 shoes.

43. An old cannon shoots 29 cannonballs in one hour. How many cannonballs can it shoot in 8 hours? *Ans:* 232 cannonballs.

44. There are 12 months in one year. How many months are in 9 years? *Ans:* 108 months.

45. There are 31 days in August. How many days are in 9 Augusts?
    Ans: 279 days.

46. There are 28 days in February, how many days are in 8 Februaries? *Ans:* 224 days.
    Solution: Solving problems with double-digit numbers ending with 8 is very much like for numbers ending with 9. Turn 28 into the closest multiplier of 10, which is 30. Multiply 30 by 8, 30 × 8 = 240. Subtract 8 twice (not once as in case of last digit being 9), 240 - 16 (two times 8) = 224 (days).

47. The width of an Olympic-size swimming pool is 82 feet. How many feet will 9 laps across the pool make?
    Ans: 738 feet (82 × 10 = 820; then, 820 - 82 = 738). By the way, the length of the Olympic pool is 164 feet.

48. Grandma takes 19 pills a day, as prescribed by her doctor. How many pills does she take in 8 days? *Ans:* 152 pills.

49. Grandpa takes 9 pills a day, as prescribed by his doctor. How many pills does he take in 31 days? *Ans:* 279 pills.

50. Grandpa exercises 58 minutes every day. How many minutes does he Exercise in one week? *Ans:* 406 minutes (60 × 7 = 420; then 420 - 14 = 406).

51. Grandma exercises for 69 minutes a day. How many minutes does she Exercise in a week? *Ans:* 483 minutes.

THE END

YOU ARE NOW READY FOR

OUR PRODUCTS
THE READING LESSON

- A step-by-step reading program for children ages 3-7

THE VERBAL MATH LESSON BOOK SERIES

- Verbal Math Lesson Level 1
  Suitable for children in K-1.

- Verbal Math Lesson Level 2
  Suitable for children in grades 1 to 2.

- Verbal Math Lesson Level 3
  Suitable for children in grades 3 to 5.

- Verbal Math Lesson Fractions
  Suitable for children in grades 5 to 7.

- Verbal Metrics - Learning Metric Conversions

- Verbal Math Lesson Percents
  Suitable for children in grades 5 to 7.

For description and prices, please see our websites:
www.readinglesson.com
www.mathlesson.com